A User's Guide to Electrical PPE

RAY A. JONES, PE

JANE G. JONES

JONES AND BARTLETT PUBLISHERS
Sudbury, Massachusetts
BOSTON   TORONTO   LONDON   SINGAPORE

Jones and Bartlett Publishers
World Headquarters
40 Tall Pine Drive
Sudbury, MA 01776
978-443-5000
info@jbpub.com
www.jbpub.com

National Fire Protection Association
1 Batterymarch Park
Quincy, MA 02169-7471
www.NFPA.org

Jones and Bartlett Publishers Canada
6339 Ormindale Way
Mississauga, Ontario L5V 1J2
Canada

Jones and Bartlett Publishers International
Barb House, Barb Mews
London W6 7PA
United Kingdom

Jones and Bartlett's books and products are available through most bookstores and online booksellers. To contact Jones and Bartlett Publishers directly, call 800-832-0034, fax 978-443-8000, or visit our website www.jbpub.com.

Substantial discounts on bulk quantities of Jones and Bartlett's publications are available to corporations, professional associations, and other qualified organizations. For details and specific discount information, contact the special sales department at Jones and Bartlett via the above contact information or send an email to specialsales@jbpub.com.

Production Credits

Chief Executive Officer: Clayton E. Jones
Chief Operating Officer: Donald W. Jones, Jr.
President, Higher Education and Professional
    Publishing: Robert W. Holland, Jr.
V.P., Sales and Marketing: William J. Kane
V.P., Production and Design: Anne Spencer
V.P., Manufacturing and Inventory Control:
    Therese Connell
Publisher, Public Safety Group: Kimberly Brophy
Publisher, Electrical: Charles Durang
Associate Managing Editor: Robyn Schafer

Production Editor: Renée Sekerak
Photo Research Manager and Photographer:
    Kimberly Potvin
Director of Marketing: Alisha Weisman
Cover Images: Imagineering
Composition: Publishers' Design and Production
    Services, Inc.
Text Printing and Binding: Malloy, Inc.
Cover Printing: Malloy, Inc.
Text and Cover Design: Anne Spencer

Library of Congress Cataloging-in-Publication Data
Jones, Ray A., P.E.
    A user's guide to electrical PPE / Ray A. Jones, Jane G. Jones.
        p. cm.
    Includes bibliographical references and index.
    ISBN-13: 978-0-7637-5114-2
    ISBN-10: 0-7637-5114-6
    1. Electric engineering—Safety measures—Handbooks, manuals, etc. 2. Protective clothing—Handbooks, manuals, etc. 3. Electrical injuries—Prevention—Handbooks, manuals, etc. I. Jones, Jane G. II. Title.
        TK152.J72 2007
        621.319'240289—dc22
        2007026593

6048

Printed in the United States of America
11 10 09 08   10 9 8 7 6 5 4 3 2

Contents

Chapter 3—Protection from Arc Flash   73

Foreword

The information in this text could save your life or the life of a coworker. Personal protective equipment (PPE) is the final barrier between you and a complete electrical circuit—sometimes a deadly enemy. You must make the decisions about what you will wear or use to keep that enemy at bay.

Consider the following account of a real electrical incident. (The names and some details have been changed to protect those involved.)

> Peter was assigned to troubleshoot the power supply on the west-end battery charger. The power supply was adjacent to the transformer that provided energy to the system. The voltmeter on the output of the battery charger read zero volts, so Peter concluded that the circuit breaker had tripped again. He grabbed his tool cart and headed to the west side of the building.
>
> Peter had worked at ProBat, Inc., for four years, and it was his first job as an electrician. The company manufactured car batteries and recently had won a major contract with two car manufacturers and an auto parts supplier. Keeping the facility operating at capacity had become a significant objective. The company developed policies to minimize expenditures. One of those policies required employees to supply their own tools. Peter purchased his voltmeter at a home improvement store when he started to work with ProBat.
>
> He had not been trained to recognize that an arc flash hazard exists in all electrical equipment. Peter had heard some rumors that OSHA was beginning to ask

employers and workers about wearing flame-resistant clothing, but he assumed that the problem was primarily associated with large manufacturing facilities and high voltage.

When Peter reached the circuit breaker in the secondary switchgear, he noticed that the 480-volt circuit breaker was still closed. He decided the problem had to be in the battery charger. He opened the door on the battery charger to see if anything unusual was visible. Everything looked normal. Peter reached for his voltmeter to check voltage at the power supply. When the probes touched the phase conductors, an explosion occurred near the voltmeter. Peter's hands were burned in the explosion.

Peter kept his voltmeter and leads lying loose in his toolbox. He considered the voltmeter to be a tool, just like his adjustable wrench, and he used it a lot. The voltmeter lead had been damaged, and eight of the nine copper strands in one of the leads had been cut, leaving a single continuous strand of the original nine strands in the conductor. A failure within the voltmeter caused the two leads to be effectively shorted together.

When Peter touched the probes to the phase conductors, he was putting a direct short circuit on the 480-volt supply conductors. Fortunately for Peter, the short-circuit current flow melted the single strand in the lead. The resulting arc expelled the molten material from the lead and cleared the fault. The arc flash energy was limited by the damaged voltmeter lead; otherwise, all the energy available in the supply transformer would change the event from burned hands to a massive arc flash event.

Peter had no PPE available to him. In the future, Peter would purchase only high-quality voltmeters. Never again would he measure voltage without wearing leather gloves. Never again would he open the door to any electrical equipment without wearing flame-resistant clothing.

ProBat changed its policy and purchased CAT IV-rated voltmeters with UL labels. ProBat now provides flame-resistant clothing for all electrical workers and trains all electrical workers to perform a hazard/risk analysis and then to select PPE based on that analysis.

This incident is one among many thousands. In most instances, the worker escapes without injury. However, a small percentage of workers are fatally injured. A higher percentage of workers are burned and permanently disfigured. Neither fatalities nor burn injuries need to occur. Both can be avoided by selection and wearing of appropriate PPE.

To those who are working tirelessly in associations, on committees, and in the industry to develop new ways to look at electrical incidents and learn from past mistakes, we say, "keep up the good work." Many people remain very concerned about the deaths and injuries that occur each year through contact with electrical equipment throughout the world. It is that concern that drives the effort to make life better for the worker.

And, finally, a special note to the manufacturers of protective equipment, clothing, and flame-resistant fabrics. You are responsible for a large part of the tremendous progress that has been made in the past few years in helping to keep workers safe from the hazards of electricity. You have been responsive to needs, you have conducted countless tests, and you have reacted positively to suggestions from the community. You have taken the lead in developing national consensus standards that guide the industry in providing a safe work place. How many lives have been saved? We will never know that number, but many workers who go home to their families each night might not have done so without your help. We thank you for providing information and photos for this book. Our original intent was

to name every manufacturer, but we quickly realized that
was an unreachable goal. The photos and names of the suppliers herein represent the entire industry, and we appreci-
ate all of you.

Ray and Jane Jones

Introduction to Electrical PPE

The technologies associated with electricity and electrical transmission are remarkable, and their invention has enabled humans to enhance their lives in many ways, from personal comfort to industrial expertise. In a perfect world, this wonderful entity called electricity would always be safe to use, but in reality, many people are hurt, maimed, and killed through its use. To take advantage of the benefits of electricity safely, individuals must use the same intelligence and skills that were used to harness electricity to address the safety issue.

Much has been done to work toward the goal of electrical safety. One tool set that workers have to protect themselves against electrical hazards is personal protective equipment (PPE). A great deal of testing has gone into the development of modern PPE, and PPE currently is available for every task. PPE is the final barrier between the worker and the hazard. It should not be the first consideration, but it should be the last consideration before a worker performs a task.

The Hazard/Risk Analysis

The most valuable safety tool available to a worker is the hazard/risk analysis. The hazard/risk analysis is the application of critical thinking to jobs or tasks that identifies whether or how a person might be exposed to hazards during the task. The analysis should identify the following:

- Whether exposure to a hazard exists
- The degree of exposure
- Whether some type of specific authorization or other control process is needed
- Whether PPE is needed
- If PPE is needed, what type of PPE will provide the necessary protection

Workers should always use a hazard/risk analysis before attempting any task in order to determine how to keep themselves safe. (More information on hazard/risk analysis can be found in the National Fire Protection Association's (NFPA) *The Electrical Safety Program Book*, available from Jones and Bartlett Publishers.)

PPE Defined

PPE stands for personal protective equipment. As the name implies, PPE provides protection to an individual. The word *personal* suggests that a person uses the equipment directly to provide protection that identifies, impedes, or eliminates his or her own exposure to a hazard. For instance, spectacles (eyeglasses) that meet the impact requirements of consensus standards should be considered personal protective equipment.

Historically speaking, to a construction worker, the term PPE means hard hat and safety glasses. To a medical technician, PPE means a face shield and latex gloves. To a hazardous waste technician, PPE means an impervious full-body suit, including a self-contained source of breathing air. To a firefighter, PPE means flame-resistant clothing, gloves, and face protection. To an electrician, PPE means voltage-rated gloves, shoes, and boots. In each instance, PPE is related to the typical hazards to which a person performing a particular set of occupational tasks is exposed. In most cases, hazards are visible and easily recognized, although one obvious exception is lack of oxygen or the existence of flammable gasses.

This book suggests that the term *electrical PPE* should be broadened to include electrical safety equipment not normally considered PPE. Certainly, PPE should be all of the equipment previously included. However, some other equipment also fits into the general description of electrical personal protective equipment. Consensus standards now include as electrical hazards fire, shock and electrocution, arc flash burns, and arc blast. Any equipment used by a person to help recognize and evaluate an electrical hazard, as well as equipment that serves as a barrier between a worker and each identified hazard, should also be considered PPE.

Products are developed and marketed with a specific purpose in mind and usually are effective for that purpose. When a product is used for multiple purposes, it cannot work as well for the other purposes as it does for the primary purpose. A product might achieve more than one objective, but the product works best when used for its intended purpose. For instance, spectacles that meet American National Standards Institute (ANSI) Standard Z87.1 (safety glasses) provide excellent protection from impact and also provide some filtering from potentially damaging ultraviolet light as a secondary purpose. The primary purpose of the spectacles is protection from impact. Thus, although the spectacles provide some protection from ultraviolet light, a worker whose primary need is protection from ultraviolet light would choose another type of spectacles. The individual must select PPE based on the hazards to which he or she is exposed.

Facilities installed as defined by installation codes (NEC® and NESC®) provide protection from fire and shock for the general public. Personal protective equipment is not needed when guards and covers are in place and equipment is functioning normally. Workers, including equipment operators, are not exposed to electrical hazards until the worker performs a task associated with the supply of electricity to the production equipment. Workers who handle portable electrical tools or equipment might be exposed to an electrical hazard if the tool or portable cord is damaged. A ground-fault circuit interrupter (GFCI) provides protection for the worker; therefore, portable GFCIs might even be considered PPE.

Electricians, technicians, and other workers are, or might be, exposed to shock and electrocution when troubleshooting a problem. To evaluate the degree of an exposure, these workers must determine the presence of and, sometimes, the level of voltage present in a circuit. Workers must use a voltage-detecting or voltage-measuring device, such as a voltmeter. A person uses these devices to determine the presence of an electrical hazard. Therefore, voltmeters and voltage detectors are PPE.

Workers use voltage-rated gloves, footwear, nonconductive head protection, and other products as barriers to direct contact with exposed energized conductors. Individuals use voltage-rated products to minimize or prevent their exposure to hazards. Blankets and mats can prevent contact or minimize exposure to an exposed energized conductor, and a person installs these products to control exposure to a shock hazard. All of these products are also PPE.

PPE also includes flame-resistant products such as clothing, coveralls, and lab coats, and arc-rated face shields and switchman's hoods used as thermal shields for protection from the elevated temperature of arcing faults. A person uses these products to control or minimize the potential for a thermal injury from an arcing fault.

Workers use live-line tools, fuse pullers, and insulated tools to prevent or minimize the risk of direct contact with an exposed energized conductor or to minimize the chance of initiating an arcing fault. A person uses these tools to control or minimize the risk of injury. These products are also PPE.

Occupational Safety and Health Administration and Electrical PPE

In 29 CFR 1910 Subpart I, Sections 132 through 139, the Occupational Safety and Health Administration (OSHA) describes expectations for personal protective equipment. Although PPE is not defined, per se, the standard suggests that an employer is required to ensure that a worker has the necessary PPE for protection from each possible hazard in the process or from all equipment associated with each active process (see 29 CFR 1910.132(a)). If workers provide their own PPE, the employer is responsible for making sure that the PPE will provide the expected protection.

In Section 1910.132(d), OSHA suggests that employers must perform a hazard assessment for the site and provide PPE for each hazard that is identified in the assessment. The section goes on to indicate that the workers must be trained to be able to select and wear PPE that will afford protection from each hazard identified in the hazard assessment. Workers must know how to inspect and maintain the PPE. Section 1910.132(f) indicates that workers must understand the limitations of specific items of PPE. OSHA requires each employer to certify that each employee required to wear PPE be trained to understand how to use the equipment. The employer is required to document the date and subject of the training for each employee.

OSHA discusses electrical PPE in Section 1910.137. The content of this section essentially mirrors consensus requirements for shock-protective equipment. Consensus standards that cover rubber-protective equipment are reviewed continually and revised frequently. Products that provide protection from electrical shock should comply with the latest revision of the appropriate consensus standard. Although Section 1910.137 covers tests and testing requirements, employers should use later revisions of the referenced standards instead of the OSHA regulations as the source of testing information.

The only electrical hazard addressed in Subpart I is electrical shock. When that standard was last promulgated (in the early 1980s), arc flash and arc blast were not recognized as electrical hazards. Therefore, this section does not cover PPE for protection from the arc flash and arc blast hazards.

PPE Ratings

PPE is designed for specific hazards. Consensus standards that define requirements for initial testing must ensure that the product provides effective protection. Ratings are established and based on the protective nature of the product. For instance, rubber products provide protection from shock and electrocution by increasing the contact resistance (or impedance) to a value that will limit current to an acceptable level. An acceptable level is determined by the circuit voltage. Therefore, rubber products are rated by voltage. Flame-resistant protective equipment is similarly rated, based on the amount of thermal insulating quality of the equipment.

In general, electrical equipment is subjected to third-party testing. However, electrical PPE normally is not third-party labeled. In the case of electrical PPE, manufacturers normally assign ratings to equipment as defined by a national consensus standard. For instance, the American Society of Testing and Materials (ASTM) D120, *Standard Specification for Rubber Insulating Gloves*, defines both standard ratings and tests necessary to verify the integrity of the product. Manufacturers use that standard as the basis for equipment rating and initial certification testing.

Flame-resistant (FR) clothing manufacturers also rely on ASTM standards to define ratings and test requirements for electrical PPE. In general, manufacturers provide a label indicating that the PPE was tested to a specific ASTM standard. However, no third-party labeling system exists for flame-resistant clothing.

A rating establishes limits beyond which the product should not be used. If a product is used beyond the rating (limit), the product probably will not provide the necessary protection. Depending on the nature of the product, the inadequately rated equipment could increase the potential for injury. The worker must not exceed any rating established either by a manufacturer or by third-party testing.

In recent years, consensus has been building about protecting workers from arc flash and arc blast. Efforts are under way to generate more understanding about how the results of an arcing fault expose a worker to thermal and pressure hazards. Until the knowledge gaps have been filled, no consensus equipment-rating system exists to help avoid an injury. However, experience provides anecdotal evidence for the protective nature of some products.

Although no method currently exists for evaluating the degree of thermal hazard at a distance closer than 18 inches from the arcing fault, experience suggests that heavy-duty leather gloves provide substantial protection from the thermal exposure. When worn in conjunction with the rubber protection in voltage-rated gloves, the combined thermal protection is very significant.

Multiple Uses

A single item of PPE possibly can provide protection from more than one hazard; however, the general rule is that PPE provides protection only from exposure to a single hazard. Until an evaluation and rating system is developed, National Fire Protection Association (NFPA) 70E advises workers to wear heavy-duty leather gloves with long gauntlets.

Workers must have the knowledge and ability to evaluate each electrical hazard in their work environment. Workers also must have the knowledge and ability to match PPE to each hazard and understand that all items of PPE have limits, and they must know what that limit is.

Relationship to Injuries

When a worker is exposed to a shock injury, essentially no room for error exists. A victim can feel current flow in the range of 0.01 ampere, which is a very small amount of current. Ventricular fibrillation could occur in that same worker if the amount of current increases to 0.1 ampere. Note that although the amount of current that can result in fibrillation is 100 times greater than the amount of current an individual can feel, fibrillation current still is only one-tenth of one ampere. PPE either prevents a fatality or it does not. There is no room for error.

FR clothing provides thermal protection from an arcing fault. Although little room for error exists to prevent a burn, burn injuries from an arcing fault are rarely fatal. However, these burns frequently are very significant and result in lengthy hospitalizations and permanent disfigurement. The amount of thermal energy that a worker receives depends on many factors, including both body position and PPE use. The critical factor is that the worker uses FR-protective equipment. Even underrated FR clothing can keep a worker's clothing from igniting and reduce extended exposure to the burning clothing.

Current methods of calculating incident energy assume that the exposed part of the worker's body is 18 inches from the arcing fault, although a worker's hand actually might be much closer. If the worker's hands are inside the restricted approach boundary, he or she should be wearing voltage-rated gloves with leather protectors. Although the leather protectors are neither intended nor rated as FR-protective equipment, the voltage-rated gloves will provide protection from very severe exposure.

Strategies for Acquiring PPE

Employers make decisions based on economics and other motives. Although motives vary from one person to another, economic choices either cost money or save money. The appropriate economic choice depends on purchasing strategies that are effective for a specific employer. In some instances, PPE is relatively inexpensive to purchase. Having the necessary equipment available when needed enables work to proceed on an orderly basis.

Many employers choose to require workers to wear FR clothing as standard dress. The FR clothing must be kept clean and maintained. If the employer chooses to purchase FR clothing for all electrical employees, the initial expenditure is significant. There are ongoing costs associated with laundering and maintaining the clothing. In some cases where an employer purchases the FR clothing for employees, the employees agree to launder and maintain the clothing until time for replacement. Both the worker and the employer are winners.

Sometimes employers choose to rent or lease PPE, especially FR clothing, from a uniform supply company. Workers are supplied with FR clothing to wear, and they return it to the clothing vendor for laundering or maintenance. Employers might find this alternative attractive. Costs for leasing or renting FR clothing vary widely from one locale to another.

When a facility has limited electrical capacity and workers do not need to wear FR protective equipment continuously, the employer might choose to purchase a limited number of uniforms or other items of FR clothing and keep them available at strategic locations on a work site. The employer must establish a process to ensure that the FR clothing is kept clean and free from damage.

■ PPE Certification and Approvals

Normally, the manufacturer establishes the rating for PPE. The manufacturer relies on national consensus standards to define what tests are necessary and then performs tests. The manufacturer may conduct performance tests internally or may utilize an external laboratory. Although formal certification may not exist, a customer may request the test data from the manufacturer.

Third-party laboratories, such as Underwriters Laboratories (UL), test and certify that specific equipment meets or exceeds defined criteria. The third-party testing system applies to equipment where the general public is or might be exposed to electrical shock. Personal protection for the general public is not a part of an "installation," and it is not covered by the scope of normally accepted installation codes for a facility.

PPE provides protection for workers who interact with energized electrical circuits and equipment, as opposed to the general public. Since PPE normally is not considered to be within the scope of installation codes, the third-party system that supports that standardization process is not used. Where third-party testing applies to PPE, the testing provides increased assurance that evaluation of the equipment is valid.

References

ANSI Z87.1, *Practice for Occupational and Educational Eye and Face Protection*. New York: American National Standards Institute, 1996.

ASTM D120, *Standard Specification for Rubber Insulating Gloves*. Conshohocken, PA: American Society of Testing and Materials, 2002.

Jones, Ray A. and Jane G. Jones, *The Electrical Safety Program Book*. Quincy, MA/Sudbury, MA: NFPA/Jones and Bartlett, 2003.

The National Electrical Code (ANSI/NFPA 70). Quincy, MA: National Fire Protection Association, 2005.

The National Electrical Safety Code (ANSI/IEEE C2). New York, NY: The Institute of Electrical and Electronics Engineers, 2007.

U.S. Department of Labor. Occupational Safety and Health Administration. *OSHA Regulations 29 CFR 1910.132-139, Subpart I, "Personal Protective Equipment."* Washington, DC.

Protection from Shock and Electrocution

Electrical shock is a common cause of electrical injuries. Electrical shock occurs when a person makes contact with an open and uninsulated energized electrical conductor (an energized circuit). No electrical current can flow unless a completed circuit exists. NFPA 70E, *Standard for Electrical Safety in the Workplace*, defines the shock hazard as "a dangerous condition associated with the possible release of energy caused by contact or approach to live parts." The worst-case scenario of shock is electrocution—death by electric shock.

Because electrical shock is caused by contact with an open and uninsulated energized conductor, the obvious solution is not to make contact with that conductor. That means that the best way to prevent shock is to ensure that the conductor is closed, insulated, and deenergized or that the conductor is so well insulated by nonconductive material that a worker can perform work safely. The best-case scenario is to deenergize the circuit; however, it is unrealistic to assume that all electrical work can be done on deenergized parts and conductors. Only specifically trained, qualified people should attempt work on energized conductors, and then only by following the detailed procedural controls of their company's electrical safety program.

■ Approach Distances

Ohm's law defines resistance, part of which is related to distance. As the worker comes closer to an open and uninsulated conductor, the chance of shock or electrocution increases. Qualified workers are more capable than unqualified workers of recognizing that the risk of injury increases with approach. As the risk of injury increases, increased controls are necessary to ensure that workers are protected from the shock and electrocution hazard.

The best way to trigger increased awareness, increased controls, and increased use of specific PPE is to define a distance from an open, energized, and uninsulated conductor and specify the PPE required for that distance. Since the shock and electrocution hazard depends on the voltage of the conductor, the distance between the energized conductor and the worker also is dependant on the voltage.

OSHA standards specify safe approach distances based on voltage. Consensus standards identify an approach distance for unqualified workers and qualified workers and also a distance that must not be crossed. The approach distance for unqualified workers is called the limited approach boundary, and the approach distance for qualified workers is called the restricted approach boundary. The boundary that must not be crossed is called the prohibited approach boundary. **Figure 2-1** illustrates the relationship of approach boundaries that are associated with electrical shock. **Table 2-1** defines the approach boundary related to shock and electrocution.

The risk of contact depends on the distance between a worker and an uninsulated energized conductor. If either the conductor or the worker is moveable, the distance will change in some circumstances. For instance, a conductor mounted on a pole line will move with the wind. A worker standing in or on a lift may not be stable. Therefore the distance between the worker and the energized conductor will change. Table 2-1 suggests a different dimension for fixed and variable distances.

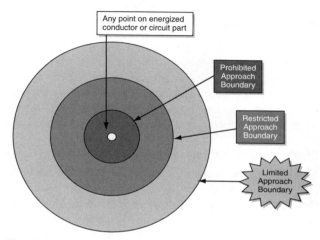

Figure 2-1 Shock Approach Boundaries

Table 2-1
Shock Hazard Boundaries

System Voltage Range, Phase to Phase	Limited Approach Boundary		Restricted Approach Boundary	Prohibited Approach Boundary
	Bus Work/ Fixed Circuit Part	Overhead Line/ Movable Conductor		
0 to 50	Not specified	Not specified	Not specified	Not specified
51 to 300	3 ft. - 6 in.	10 ft. - 0 in.	Avoid contact	Avoid Contact
301 to 750	3 ft. - 6 in.	10 ft. - 0 in.	1 ft. - 0 in.	0 ft. - 1 in.
751 to 15 kV	5 ft. - 0 in.	10 ft. - 0 in.	2 ft. - 2 in.	0 ft. - 7 in.
15.1 kV to 36 kV	6 ft. - 0 in.	10 ft. - 0 in.	2 ft. - 7 in.	0 ft. - 10 in.
36.1 kV to 46 kV	8 ft. - 0 in.	10 ft. - 0 in.	2 ft. - 9 in.	1 ft. - 5 in.
46.1 kV to 72.5 kV	8 ft. - 0 in.	10 ft. - 0 in.	3 ft. - 3 in.	2 ft. - 1 in.
72.6 kV to 121 kV	8 ft. - 0 in.	10 ft. - 0 in.	3 ft. - 2 in.	2 ft. - 8 in.

Note: The appropriate flash protection boundaries that require PPE must be observed.
Source: NFPA 70E

■ The Shock and Electrocution Hazard

Electrical current flowing through a copper conductor has multiple results. The impact of current flowing through the resistance of the conductor generates heat. When the current reaches the intended equipment, the current then produces the expected effect, such as causing a motor to turn or move or producing heat from a heating element.

Ohm's Law and the Human Body

Electrical current flowing through tissue in the human body also has multiple results. Just as a copper wire resists the flow of current, tissue in the human body resists the flow of electrical current. The relationship of current flow to resistance is known as Ohm's law, named after the German physicist Georg Simon Ohm who discovered this principle.

At a specific voltage, resistance determines the amount of current flow (for direct current) or impedance (in alternating current) of the current path. Ohm's law says that at a specific voltage, volts (V) equal the amount of current flow in amperes (I) multiplied by the resistance or impedance of the circuit (R) in ohms:

$$V = I \times R$$

Current (I) is what flows on a wire or conductor, like water flowing down a river. Current flows from points of high voltage to points of low voltage in a conductor. Current is measured in amperes or amps.

Voltage (V) is the difference in electrical potential between two points in a circuit. It is the push or pressure behind current flow through a circuit, measured in volts.

Resistance (R) determines how much current will flow through a component. A very high resistance allows a small amount of current to flow. A very low resistance allows a large amount of current to flow. Resistance is measured in ohms.

Voltage/Potential

Sometimes voltage is called potential. Voltage is the force that drives current through a conductor. When a conductor is at one voltage and a second conductor is at a different voltage, the voltage difference between the two conductors is called potential difference. When a circuit is initiated between the two conductors, the resistance or impedance of the path that completes the circuit determines the amount of current that will flow in the circuit. Ohm's law determines the amount of current.

Resistance

Resistance is the physical characteristic of a material that is determined by the ability of the material to release electrons (or permit current to flow in a direct-current (DC) circuit). When the source of energy is alternating current (AC), inductance and capacitance affect the ability of the circuit to conduct current. The combination of resistance, inductance, and capacitance is called impedance. Therefore, resistance and impedance have similar impact on current flow, depending on whether the voltage source is direct or alternating.

Resistance to current flow in a conductor is manifested as heat. If resistance remains the same and the amount of current increases, the amount of heat increases. The copper conductor must dissipate the heat, or the conductor will overheat and melt. In a copper conductor, the ability of the conductor to dissipate heat determines the current or ampacity rating of the conductor. To reduce the heat generated by current flow, the resistance (impedance) will be lowered by increasing the size of the conductor.

However, if the conductor is human tissue, heat from current flow results in elevating the temperature of the tissue, which is already at about 98°F. In a human body, the size of the conductor (such as an arm, hand, or finger) is fixed. Therefore, the resistance (impedance) of the circuit is fixed. The amount of current is directly proportional to the amount of voltage. If the voltage is increased, the current flow also will increase. The circuit voltage, then, is a critical characteristic to determine the amount of current flow, which generates heat energy in the tissue.

Different conductor materials resist the flow of current at different rates. For example, the impedance of steel is much greater than the impedance of aluminum. The impedance of aluminum is greater than the impedance of copper. Likewise, different tissues in the human body have different impedance values. For example, nerve tissue is a better conductor than skin tissue. Blood is essentially an electrolyte and is, therefore, a good conductor. Together, different body tissues provide parallel circuits and define overall body impedance.

When direct contact is made between a person and an energized conductor, current (as defined by Ohm's law) flows. The resistance or impedance of the human circuit comprises two components: the internal impedance, plus the impedance of the contact area. The majority of the total impedance of the human circuit is the contact impedance. If the contact surface is dry, the impedance is high. If the surface of either the conductor or the skin surface is wet (for example, from perspiration), the impedance is lower. If the skin surface is damaged by a cut or blister, the contact impedance is reduced even further.

Current Flow in the Human Body

Normally in the human body, a signal (voltage) is chemically generated within the central nervous system. The voltage generated within the central nervous system results in current flow through nerve tissue to a specific muscle. The muscle then moves as directed by the signal. The amount of voltage (signal) is in the millivolt or microvolt range. When a muscle receives a signal, the muscle contracts and remains contracted until the signal is removed. The muscle then releases. When a human makes contact with an energized conductor, current from the external source of voltage causes nerves to conduct a strong signal to various muscles. Muscles respond by contracting, and the body becomes rigid. If the person's hand happens to contract around the energized conductor, the person probably is unable to release and move the hand. The point where the person cannot release the hand is called the let-go threshold (see **Table 2-2**). When the current flow reaches or

Table 2-2
Reaction of Human Body to Electrical Current

Effect of Current	* AC Current in Amperes—Men	* AC Current in Amperes—Women
Perception threshold (tingling sensation)	0.0010	0.0007
Slight shock—not painful (no loss of muscle control)	0.0018	0.0012
Shock—painful (no loss of muscle control)	0.0090	0.0060
Shock—severe (muscle control loss, breathing difficulty)—onset of "let-go" threshold	0.0230	0.0150
Possible ventricular fibrillation (3-second shock)	0.1000	0.1000
Possible ventricular fibrillation (1-second shock)	0.2000	0.2000
Heart muscle activity ceases	0.5000	0.5000
Tissue and organs burn	1.5000	1.5000

*Because women's frames and body parts, such as hands and fingers, are often smaller than men's, women tend to suffer damage at lower amounts of current density or exposure.
Source: U.S. Department of Energy Electrical Safety Guidelines

exceeds the let-go threshold, the person cannot release the grip.

Current flow then continues to increase until all of the muscles in the person's body are unable to function. The diaphragm is a muscle that functions automatically, unless the muscle is instructed by an external source to contract and cause breathing to stop. Only another external force, such as a person, can remove the source of current or break the contact with the energized conductor. A person can live only for a few minutes without breathing.

The heart functions similarly. When the heart muscle remains contracted, it can no longer pump blood through the arteries and veins. Various organs within the body fail from the lack of oxygen. Death is almost guaranteed.

To prevent electrocution, limiting the amount of current flowing in tissue is essential. The voltage of the circuit cannot be changed if the circuit cannot be deenergized. According to Ohm's law, only two characteristics affect the amount of current flow in a circuit: circuit voltage and total circuit impedance (resistance). Since internal body resistance cannot be changed, contact resistance is the only remaining characteristic that might be modified.

For current to flow in the human body, the person must simultaneously contact both an exposed conductor energized at one potential and another conductor at a different potential so that the body experiences a difference of potential. To effectively eliminate the chance of current flowing within a body, contact with either or both potentials must be eliminated. The contact resistance at either conductor can be increased to a safe level by inserting an insulating material between the person and either or both conductors.

In most instances one conductor is a grounded enclosure, a grounded conductor, or earth. In general, earth serves as a reference point for an electrical circuit and is considered to be at a potential of zero volts. Note that concrete and other walking surfaces are conductive and can act as one conductor in an electrical circuit where a shock or electrocution occurs.

Insulation and the Human Body

If the person is insulated from earth ground, one possible completed circuit is eliminated. If a person is insulated from an energized conductor, another possible completed circuit is eliminated. If a person uses insulated tools, still another possibility is eliminated. Personal protective equipment was invented to address all of these possibilities.

Insulating mats, footwear, and other barriers increase the impedance of the human circuit. Insulating blankets and line hose increase the impedance between a person and the exposed energized conductor. Insulating gloves and sleeves increase the impedance between a person and the exposed energized conductor. Hot sticks and insulated tools provide insulation between a person and the exposed energized conductor.

Note, however, that insulated and insulating equipment only changes or modifies the impedance of the human circuit. This equipment is not intended to provide protection from any electrical hazard other than shock. When a worker performs a hazard/risk analysis, therefore, he or she must consider all hazards, including arc flash and arc blast.

▐ Ratings and Standards

To protect from shock or electrocution, workers must have a method to select personal protective equipment that will provide the needed increase in impedance. The rating system must include a process to verify that the PPE continues to provide the necessary impedance as the equipment is used in the field. Current codes and standards define a system that provides assurance that PPE can provide the necessary increase in impedance.

The American Society of Testing and Materials (ASTM) writes and publishes voluntary standards that define characteristics of insulated and insulating products based on circuit voltage. ASTM has assumed responsibility of writing standards for manufacturing, testing, and maintaining standards for electrical insulating products. The ASTM committee is called the F18 Electrical Protective Equipment for Workers committee. The F18 committee has jurisdiction over electrical protective equipment. Each product within its jurisdiction is assigned to a subcommittee with direct responsibility for the content of the standard, including its effectiveness in accomplishing the goal stated in the subcommittee scope.

A subcommittee (F18.15) of the F18 committee of electrical protective equipment for workers has jurisdiction over gloves and protectors, insulating sleeves, climbing equipment, and footwear. The F18.15 subcommittee produces standards that define all aspects of these products. ASTM standards are intended to be voluntary.

Although the standards are voluntary, OSHA utilizes the ASTM standards on electrical personal protective equipment. In 29 CFR 1910.137, which addresses electrical protective equipment, OSHA suggests that the current OSHA rules are based on previously published editions of specific ASTM standards. In a note that follows paragraph 1910.137 (a)(3)(ii)(B), OSHA indicates that implementing requirements in current and future editions of the standards is deemed to be in compliance with the requirements of 1910.137.

The American National Standards Institute (ANSI), which is the governing body for U.S. participation in international standards efforts, has adopted many of these

ASTM standards as well as standards produced by other standards-developing organizations (SDOs). The standards then are published as American National Standards. ANSI appoints its committees to act as secretariat and to draft some standards that address safety concerns, such as spectacles and hard hats. ANSI standards are recognized legally as American National Standards. For ASTM and ANSI standards that cover shock protective PPE, see **Table 2-3**.

Mitigating the Hazard

One of the ways that workers can mitigate, or lessen, the shock hazard is by using PPE. However, use of PPE is not the first or even the best way to keep people from being shocked or electrocuted. Using PPE is the last line of defense against electrical hazards. It is the final barrier between a person and the electrical hazard. That final barrier exists only if the person chooses appropriate equipment and uses it or wears it.

PPE is commonly thought to mean protective clothing or apparel—safety glasses and shoes, insulated gloves, or fire-retardant clothing. It is that, but it is also protective equipment: rubber blankets, ground-fault circuit interrupters (GFCIs), voltmeters, safety labels, and insulated hand tools. Even electrical safety procedures and standards can be considered PPE.

Companies that expect their qualified workers to perform work on energized parts and conductors (live parts) must provide the information and training that the workers need to protect themselves. PPE must be selected carefully for each job and must be maintained in perfect condition, as lives depend upon it.

It is unrealistic to expect that all grounded components are isolated from contact with the worker. Since contact with earth ground generally is one electrode in the current

Table 2-3
Standards for Shock Protection

Subject	Number and Title
Gloves and sleeves	ASTM D120, *Standard Specification for Rubber Insulating Gloves*
	ASTM D1051, *Standard Specification for Rubber Insulating Sleeves*
	ASTM F496, *Standard Specification for In-Service Care of Insulating Gloves and Sleeves*
	ASTM F696, *Standard Specification for Leather Protectors for Rubber Insulating Gloves and Mittens*
Footwear	ASTM F1116, *Standard Test Method for Determining Dielectric Strength of Dielectric Footwear*
	ASTM F1117, *Standard Specification for Dielectric Footwear*
	ASTM F2413, *Standard Specification for Performance Required for Foot Protection*
	ANSI Z41, *Standard for Personal Protection-Protective Footwear*
Inspection of protective rubber products	ASTM F1236, *Standard Guide for Visual Inspection of Electrical Protective Rubber Products*
Blankets and matting	ASTM D178, *Standard Specification for Rubber Insulating Matting*
	ASTM D1048, *Standard Specification for Rubber Insulating Blankets*
	ASTM F479, *Standard Specification for In-Service Care of Insulating Blankets*
Line hoses and covers	ASTM D1049, *Standard Specification for Rubber Covers*
	ASTM D1050, *Standard Specification for Rubber-Insulating Line Hoses*
	ASTM F478, *Standard Specification for In-Service Care of Insulating Line Hoses and Covers*

Table 2-3
(Continued)

Subject	Number and Title
Ladders	ANSI A14.1, *Safety Requirements for Portable Wood Ladders*
	ANSI A14.3, *Safety Requirements for Fixed Ladders*
	ANSI A14.4, *Safety Requirements for Job-Made Ladders*
	ANSI A14.5, *Safety Requirements for Portable Reinforced Plastic Ladders*
Fiberglass tools/ladders	ASTM F711, *Standard Specification for Fiberglass-Reinforced Plastic (FRP) Rod and Tube Used in Live-Line Tools*
Plastic guards	ASTM F712, *Standard Test Methods for Electrically Insulating Plastic Guard Equipment for Protection of Workers*
	ASTM F968, *Standard Specification for Electrically Insulating Plastic Guard Equipment for Protection of Workers*
Voltage detectors	ASTM F1796, *Standard Specification for High-Voltage Detectors-Part 1 Capacitive Type to Be Used for Voltages Exceeding 600 Volts AC*
Live-line tool grounds and bypass jumpers	ASTM F711, *Standard Specification for Fiberglass-Reinforced Plastic (FRP) Rod and Tube Used in Live-Line Tools*
	ASTM F855, *Standard Specification for Temporary Protective Grounds to Be Used on De-energized Electrical Power Lines and Equipment*
	ASTM F2249, *Standard Specification for In-Service Test Methods for Temporary Grounding Jumper Assemblies Used on De-Energized Electric Power Lines and Equipment*
	ASTM F2321, *Standard Specification for Flexible Insulated Temporary By-Pass Jumpers*
Insulated hand tools	ASTM F1505, *Standard Specification for Insulated and Insulating Hand Tools*

ANSI–American National Standards Institute
ASTM–American Society for Testing and Materials

flow circuit, the worker must pay attention to the position of his or her body to ensure minimum exposure to simultaneous contact with an energized component and earth ground.

Types of Protection

Several types of PPE are necessary to help mitigate the electrical shock hazard. This chapter discusses the following types:

- Gloves
- Sleeves
- Mats
- Blankets
- Line hose
- Footwear
- Live-line tools (hot sticks)
- Gloves

Gloves

Gloves increase the resistance of the current path between an energized electrical conductor and a person's skin. By increasing the impedance, the gloves reduce the amount of current that flows through the person's body as a result of a contact (accidental or intentional). The amount of current that flows through body tissue is directly proportional to the kind and degree of injury.

If the insulation value of a worker's glove is overcome by the voltage potential of the electrical conductor, the glove resistance decreases to the point that dangerous currents might flow through the worker's body. Workers must have the ability to select an insulating glove that cannot be affected by the voltage potential of the circuit.

Hazards

Voltage-rated gloves are intended to protect a worker from electrical shock or electrocution. By increasing the imped- ance or resistance of the potential current path that in- cludes a worker's body, the amount of current flow is reduced to an imperceptible level. The risk of electrocution is controlled. Rubber gloves provide the increased imped- ance or resistance, and the leather protectors help to elimi- nate the chance that the rubber insulating layer can be penetrated while the task is being executed.

The rubber used in the insulating layer can burn, but it is difficult to ignite. The leather protectors offer signifi- cant protection from the thermal hazard associated with an arcing fault. Although leather protectors have no specific incident energy rating, experience shows that they offer significant protection from exposure to an arcing fault (see **Figure 2-2**). A subcommittee (F18.15) of the F18 commit- tee of electrical protective equipment for workers has juris- diction over gloves and protectors, insulating sleeves, climbing equipment, and footwear. The F18.15 subcom- mittee produces standards that define all aspects of these products.

Degradation

Ozone is a naturally occurring form of oxygen. Near the earth's surface, ozone molecules combine easily with other ozone molecules to form oxygen. Ozone is generated and frequently exists in the vicinity of electrical arcs that occur during switching operations and during fault conditions. Gloves manufactured of natural rubber degrade over time when exposed to ozone. Gloves made from this compound are designated Type I. Currently, very few Type I gloves are manufactured or used because of the degradation of the material by exposure to ozone or ultraviolet energy.

Figure 2-2   Gloves
Courtesy of Salisbury Electrical Safety LLC

Gloves manufactured from certain elastomers are resistant to ozone degradation and are designated as Type II. Gloves constructed from these elastomers are susceptible to damage from abrading, puncturing, and cutting.

A worker's hands tend to perspire when wearing rubber gloves. Hands that are slightly moist from perspiration do not slide easily into the rubber gloves. Many workers sprinkle powder into the glove before sliding their hand into them. Historically, workers have used talcum power for that purpose. That practice should be avoided, because talc hastens the degradation of the rubber-insulating compound. Two alternatives are available:

1. The worker can use a powder provided by the manufacturer that contains no talc or other product that will degrade the insulating material.

2. The worker can use liners that absorb perspiration, thus keeping the worker's hands dry and enabling the rubber gloves to slide on and off easily.

Voltage Rating

In addition to being designated as Type I or Type II, insulating gloves are assigned to a class that is dependant on the maximum voltage on which the gloves may be used safely. As might be expected, as the designated voltage level increases, the thickness of the insulating compound also increases. Having a thicker insulating layer makes the gloves less pliable and less flexible.

Gloves are assigned a voltage rating that corresponds to the maximum circuit voltage at which the glove can be used safely. Insulating gloves are designated as Class 00, Class 0, Class 1, Class 2, Class 3, and Class 4. As the class increases, the safe voltage also increases. **Table 2-4** indicates the upper voltage limit on which the gloves may be used.

Table 2-4
Rubber Glove Labeling Chart

Class Color	Label Color	Conventional Work Position for Worker	Voltage Rating
00	Beige	Ground, structure, or basket	500 / 750 (dc)
0	Red	Ground, structure, or basket	1000 (ac) / 1500 (dc)
1	White	Structure or basket	7500 (ac) / 11,250 (dc)
2	Yellow	Electrically isolated basket or platform	17,000 (ac) / 25,500 (dc)
3	Green	Electrically isolated basket or platform	26,500 (ac) / 39,750 (dc)
4	Orange	Electrically isolated basket or platform	36,000 (ac) / 54,000 (dc)
Source: Salisbury Electrical Safety LLC			

Voltage-rated gloves may be used on conductors lower than their rating, but not higher than the glove rating. For instance, Class 1 gloves (7,500-volt rating) may be used on a 1,000-volt circuit; however, Class 0 gloves (1,000-volt rating) may not be used on 5,000-volt circuit. The voltage rating of the glove must be at least equal to or higher than the applicable voltage.

Size and Style

Voltage-rated gloves should fit the hands of the person wearing them. Glove size for a specific hand is determined by measuring the circumference of the palm at its widest point. The palm measurement is the size that should be ordered. **Figure 2-3** illustrates how to determine glove size.

Voltage-rated gloves are available with multiple cuff designs. The default design is the straight cuff. Other cuff designs have advantages in specific circumstances. For

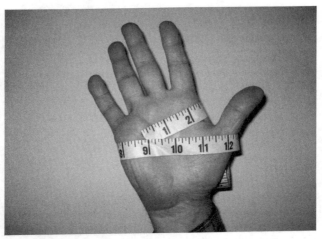

Figure 2-3   Measuring the Hand for Glove Fit

instance, bell cuff or flare cuff designs accommodate heavier clothing that might be required during winter months.

Leather Protectors

Because the purpose of voltage-rated gloves is to provide the insulating qualities necessary to reduce current flow to an imperceptible level, the rubber must be protected from damage in use. Even a pinhole in the rubber could result in an electrocution. Abrasion from inadequate protection while the gloves are in storage or in use could damage the rubber and result in an electrocution. To avoid these possibilities, protectors must be worn over the rubber insulating gloves, and the gloves should be stored assembled with their protectors (see **Figure 2-4**). Although the added layer of leather reduces the flexibility of the gloves, the extra

Figure 2-4   Leather Protectors
Courtesy of Salisbury Electrical Safety LLC

layer is necessary to avoid damage to the insulating value of the rubber. Manufacturers who supply insulating gloves also provide leather protectors for each class of gloves.

The function of leather protectors, sometimes called leathers, is to reduce the chance of damage to the insulating glove. As the circuit voltage increases, circuit components tend to be larger and heavier. The larger, heavier components suggest that the leather protection needs to be more rugged.

A worker's hand and arm are in contact with the interior surface of the gloves while he or she is wearing them. The exterior surface of the insulating rubber is in contact with the leather protector. Although the rubber insulating material is tested and rated to adequately protect a worker from shock, the material in the leather protector has no such characteristic. When a worker touches an energized conductor with the gloves, the entire leather protector is energized at the same voltage of the conductor. Therefore, the leather protector must not contact any uninsulated body part. The insulating rubber glove must extend beyond the leather protector by a length that is sufficient to eliminate the chance of creepage between the leather protector and the worker's skin. Gloves insulate the hands of the worker from current flow (shock) because the rubber is an insulator. Leather is not an insulator; therefore, leather protectors protect only the gloves, not the hands. In an arc flash, however, leather protectors become the protecting component (see Chapter 3).

Selection and Use of Voltage-Rated Gloves
Rubber gloves used for shock protection should be called rated gloves or voltage-rated gloves. This language provides the best chance to avoid miscommunication about hand protection. Manufacturers produce and market several different types of rubber gloves. These products have impor-

tant uses, but the uses do not include insulation from energized electrical conductors. Although many different types of rubber gloves are on the market, unless they meet consensus requirements as voltage-rated gloves, they should never be used for shock protection (see **Figure 2-5**).

Voltage-rated gloves must be worn any time a worker crosses the prohibited approach boundary. (Note that administrative controls such as an energized work permit should apply if the worker must cross the prohibited approach boundary.) Workers who must work within the restricted approach boundary should consider using voltage-rated gloves for protection from unintentional contact with an energized conductor.

Figure 2-5   Voltage-Rated Gloves
Courtesy of Salisbury Electrical Safety LLC

Gloves that have the lowest voltage rating that exceeds the circuit voltage should be selected and worn. As the voltage rating increases, the flexibility of the gloves decreases. Therefore, maximum flexibility is achieved by using gloves with the rating nearest that of the circuit voltage.

Purchase

ASTM D120, *Standard Specification for Rubber Insulating Gloves*, is the national consensus standard that defines the physical characteristics of the gloves. The standard defines the minimum electrical, chemical, and physical properties of voltage-rated gloves. Purchase orders should require products that comply with ASTM D120. In addition to that basic requirement, the purchase order should also specify the following information:

- Type
- Class
- Length
- Size
- Color
- Cuff design

Gloves that comply with ASTM D120 have a colored label that provides the required information. The color of the label corresponds to the voltage rating of the glove. Although the gloves may be different colors, ASTM D120 defines the label color (see Table 2-4). Voltage-rated gloves should always be purchased with leather protectors and a storage container.

Storage

Voltage-rated gloves should always be kept in the storage container when not in use. The insulating integrity of the construction material must be maintained. Gloves, assembled with their leather protectors, should be kept in a storage container with the fingers pointing toward the opening of the storage container. The storage container should be kept in a clean and dry location when not in use.

Inspection

Gloves and leather protectors must be inspected before each use. Although nationally recognized standards require visual examination and inspection of both the insulating glove and the leather protector before use, the real reason for inspecting them is to prevent electrocution. Any damage to either the insulating material or the leather protectors can result in the wearer being electrocuted when direct contact is made with an exposed energized conductor.

After removing the gloves from the storage container, the rubber insulating gloves should be removed from the leather protectors. The first inspection point is to verify that the voltage rating of the gloves is at least as great as the maximum voltage associated with the work task.

All surfaces of the leather protectors should be visually inspected for cuts and abraded surfaces and should be replaced if any deficiency is detected. The purpose of inspecting the leather protectors is to locate deficiencies that might result in damage to the insulating rubber. Leather protectors are easily damaged when used with large conductors, and such damage may be sufficient reason to replace the set of gloves. The surface of the fingers and palm area of the leather protectors should be inspected carefully to find any place where shards of wire might have penetrated the leather. Of course, should a deficiency be identified, the gloves should not be used. The fingers should be cut off the gloves to ensure that no one will inadvertently use them, and then they should be discarded. The leather protectors should also be inspected for excess dirt and lubricant. Leather protectors contaminated with petroleum-based lubricants should be discarded.

All surfaces of the rubber-insulating glove should be inspected for any indication that the insulating characteristics might be reduced. The inspection should focus on the fingers and palm area of the glove and any potential cuts or abrasions to the surface of the insulating material (see **Figure 2-6**). The rubber insulating material can be damaged from splinters on wood poles, sharp corners and edges on metal enclosures, and many other sources. The inspection must ensure that the insulating glove has not sustained such damage. If a damaged area is not found before the gloves are used, the wearer could be electrocuted.

The worker must inspect the gloves for damage from ultraviolet (UV) energy by rolling the glove surface between

Figure 2-6   Damaged Gloves
Courtesy of Salisbury Electrical Safety LLC

the forefinger and thumb. The worker must inspect the surface of the glove carefully to make sure it is smooth as the rubber is rolled between the finger and thumb. If the rubber appears to be strained, the insulating integrity of the rubber may have been breached by UV or chemical damage. If the surface of the rubber glove is not smooth, the glove should be removed from service and replaced. If deterioration is found on either glove, both gloves should be removed from service and destroyed. Workers should avoid folding the rubber gloves during the inspection. Folding places unnecessary strain on the insulating material and should be avoided.

After completing the visual inspection, the next step is to locate pinholes in the insulating material. Workers should inflate the rubber gloves and listen for air escaping through a hole in the insulating layer. Gloves may be inflated with a manual air pump or by holding a glove at the opening in the cuff and quickly swinging the glove to trap air inside. The air is held inside the glove and the glove is expanded by forcing the trapped air into a smaller and smaller region. The gloves should not be overinflated, as this may stress and damage the insulating material (see **Figure 2-7**).

This inspection should be performed on both gloves. If any deficiency is identified, the fingers of the gloves should be cut off to prevent their use. For a flow chart for glove inspection, see **Figure 2-8** .

Testing

When voltage-rated gloves are issued to a worker, the worker should remove the gloves from the storage container and immediately look for the label indicating the next test date for the gloves. Voltage-rated gloves must be subjected to routine tests to ensure that the insulating integrity of the gloves is maintained. Gloves must be tested as

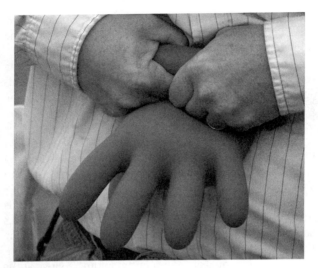

Figure 2-7   Glove Inspection
Courtesy of Salisbury Electrical Safety LLC

defined in ASTM F496, *Specification for In-Service Care of In-sulating Gloves and Sleeves*. This standard describes the process used to determine whether the gloves can protect a user. The integrity of voltage-rated gloves must be verified at intervals not exceeding six months, and employers must certify that the voltage-rated gloves have been tested within the previous six months. The certification may take the form of employer-maintained dated inspection records, or the employer may ensure that the date of the last test was marked on the cuff of the glove in non-conductive ink by the testing lab.

When new voltage-rated gloves are purchased, the gloves should be tested and prepared for use. New or recer-tified voltage-rated gloves may be kept in storage for up to 12 months before being issued. The new gloves then may be used for up to six months before they are required to be

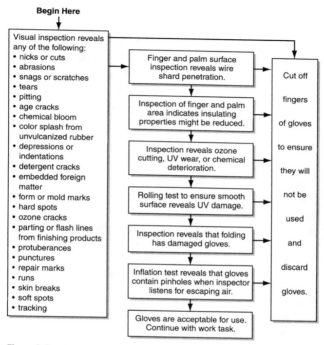

Begin Here

Visual inspection reveals any of the following:	

- nicks or cuts
- abrasions
- snags or scratches
- tears
- pitting
- age cracks
- chemical bloom
- color splash from unvulcanized rubber
- depressions or indentations
- detergent cracks
- embedded foreign matter
- form or mold marks
- hard spots
- ozone cracks
- parting or flash lines from finishing products
- protuberances
- punctures
- repair marks
- runs
- skin breaks
- soft spots
- tracking

Finger and palm surface inspection reveals wire shard penetration.

Inspection of finger and palm area indicates insulating properties might be reduced.

Inspection reveals ozone cutting, UV wear, or chemical deterioration.

Rolling test to ensure smooth surface reveals UV damage.

Inspection reveals that folding has damaged gloves.

Inflation test reveals that gloves contain pinholes when inspector listens for escaping air.

Gloves are acceptable for use. Continue with work task.

Cut off fingers of gloves to ensure they will not be used and discard gloves.

Figure 2-8   Glove Inspection Flow Chart

retested. However, gloves that have been issued and used must be retested at intervals not exceeding six months.

The process for retesting voltage-rated gloves is defined by ASTM F496. The process also is described in OSHA 29 CFR 1910.137, which addresses electrical protective equipment.

Regulatory and consensus standards assign responsibility to employers to do the following:

- Supply the necessary voltage-rated gloves.
- Maintain the protective nature of voltage-rated gloves.

- Ensure that the voltage-rated gloves are available when needed.

However, NFPA 70E, *Standard for Electrical Safety in the Workplace*, assigns workers the responsibility of implementing the requirements defined by the employer. Employers and employees alike must accept their roles in eliminating injuries.

Sleeves

Sleeves serve to increase the impedance of the current path between an exposed energized conductor and a person attempting to manipulate the conductor or another circuit part. By increasing the impedance, the sleeves reduce the amount of current flow to a predetermined level. The objective, of course, is to reduce the current flow to a level that cannot harm the person wearing the sleeves. Sleeves must never be worn without voltage-rated gloves.

Hazards

Sleeves are considered PPE that can protect a worker from electrocution. By limiting the amount of current flow through a worker's body, the chance of electrocution is effectively eliminated. Sleeves are not intended to protect a worker from the thermal hazard associated with an arcing fault. The product is constructed from natural or synthetic rubber and will burn. Should the rubber ignite, significant fuel load would be available and the worker is likely to receive a very severe burn. If a worker is exposed to the thermal hazard in addition to electrical shock or electrocution, no effective protective equipment exists. The only viable alternative is to remove the exposure by deenergizing the conductor.

Construction

Sleeves are assigned to the same voltage classes as voltage-rated gloves (see **Table 2-5**). However, sleeves are not avail-

Table 2-5
Classes of Voltage-Rated Sleeves

Class Color	Label Color	Voltage Rating
0	Red	1000 (ac) / 1500 (dc)
1	White	7500 (ac) / 11,250 (dc)
2	Yellow	17,000 (ac) / 25,500 (dc)
3	Green	26,500 (ac) / 39,750 (dc)
4	Orange	36,000 (ac) / 54,000 (dc)
Source: ASTM D120, Salisbury Electrical Safety LLC		

able for class 00. Sleeves are manufactured from natural rubber and from thermoplastic materials. Natural rubber sleeves are identified as Type I, and thermoplastic sleeves are identified as Type II. These are the same designations assigned to voltage-rated gloves. (See **Table 2-6** for types of sleeves that are available.)

Since Type I sleeves are constructed from natural rubber, they degrade over time when exposed to ozone that exists naturally near the earth's surface. Ozone normally is generated in the vicinity of an electrical arc that occurs during normal switching operations or during an electrical fault. Ultraviolet radiation increases the rate at which Type I sleeves degrade over time. Type I sleeves should be protected from such exposure to ozone and ultraviolet radiation when not in use. Type II sleeves are resistant to ozone checking and damage from ultraviolet radiation exposure.

Sleeves are manufactured by dipping or by molding. In the dipping process, a mold is dipped into the liquid insulating material until the desired thickness is achieved. Injecting or compressing the insulating material into a mold produces molded sleeves.

Type I sleeves are available as straight taper (style A) or curved elbow (style B). Some manufacturers provide

Table 2-6
Types of Sleeves

Process	Voltage Class	Color	Material
Dipped	0	Yellow	Rubber
	1	Red	Rubber
	2	Black	Rubber
	3	Yellow inside, black outside	Rubber
	4	Red inside, black outside Yellow inside, red outside	Rubber
Molded Type I	1	Black	Rubber
	2	Black, yellow	Rubber
	3	Yellow	Rubber
	4	Maroon	Rubber
Molded Type II rubber	1	Black, orange	Synthetic
	2	Black, orange	Synthetic rubber
Source: Salisbury Electrical Safety LLC			

sleeves with extra curving in the elbow. A worker should select the construction that most nearly fits the arm position when the work task is performed. Type II sleeves are available in the curved construction only. The curved style most nearly fits the needs of all work tasks that require sleeves.

When sleeves are worn, the end of the sleeve must reach inside the cuff of each voltage-rated glove. The sleeves must have the same (or greater) voltage rating as the voltage-rated gloves. Sleeves must never be worn alone or on a single arm. They must always be worn in pairs and cover both arms from below the cuff of each glove to the shoulder of the person wearing the sleeves (see color insert, **Figure CP-1**).

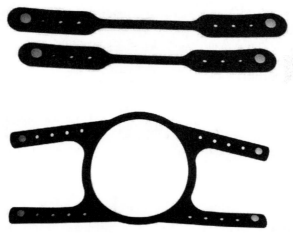

Figure 2-9   Straps and Harness
Courtesy of Salisbury Electrical Safety LLC

Sleeves are held in place by buttons and straps or a harness (see **Figure 2-9**). The buttons and straps are not intended to provide any protection from current flow. The sole purpose of these accessories is to hold the sleeves in place on the worker's arm. Both Type I and Type II sleeves are constructed with holes to accept buttons with which to fasten the straps to the sleeves.

Selection and Use

Before beginning any work task, the worker should perform a hazard/risk analysis. The analysis should consider if the worker's arm could penetrate the prohibited approach boundary while executing the task. When a worker's arm could penetrate the prohibited approach boundary, the worker should wear sleeves with the appropriate voltage rating. Of course, the best alternative is to deenergize the circuit before executing the work task. However, the next best alternative is to ensure that all body parts that will be within the restricted approach boundary are protected from potential contact with exposed energized conductors.

Rubber insulating sleeves must be tested as defined in ASTM D1051, *Standard Specification for Rubber Insulating Sleeves*, before the first use and then annually. The testing lab marks the date of the latest test on each sleeve with nonconductive ink. Although not required, sleeves should be tested at intervals of six months. Testing sleeves and gloves at the same time simplifies the administrative process.

Purchase

Purchase orders for sleeves should contain a requirement for the product to comply with all requirements defined in ASTM D1051. The following information must be selected and provided on the purchase order:

- Voltage class
- Right hand or left hand
- Type I or Type II
- Size
- Construction—molded or dipped
- Color
- Style—straight (style A) or curved (style B)

The purchase order also must define required accessories, such as the following:

- Harness (one required per pair of sleeves)
- Buttons (describe one-piece or two-piece button—four required for each pair of sleeves)
- Straps with buttons (describe strap length of either 13.5 or 15 inches—two required for each pair of sleeves)

Storage

Insulating sleeves should be stored in a container specifically designed to help maintain the integrity of the insulating material. The container with the sleeves should be stored in a clean, dry area. The sleeves should fit in the storage container without imposing strain or stress on the insulating material. The sleeve should never be folded or forced into a storage container that is too small.

Inspection

Sleeves must be visually inspected before each use. First, verify that the voltage rating of the sleeves is at least as great as the voltage of the circuit where the work task will be performed. In addition, the voltage rating of the sleeves should match the voltage rating of the companion gloves.

The complete surface of the sleeves must be inspected carefully for cuts, abrasion, ozone or chemical damage, penetration by a foreign object, or any other sign of potential damage. If any indication of damage that could affect the insulating integrity is observed, the sleeves must not be used. The physical inspection should be particularly intense near the elbow or forearm area of the sleeve. Both Type I and Type II sleeves are susceptible to penetration by the sharp ends of wire, as only a single strand of wire can easily penetrate the insulating material.

If a deficiency is found or suspected, the sleeves should not be worn. Although consensus standards permit repair of some deficiencies, repair is not recommended. The sleeves should be destroyed and then discarded.

Mats and Matting

A rubber mat is essentially a subset of matting (see **Figure 2-10**). Mats are intended to be installed on the floor in front of electrical equipment. The mat surface may be smooth, corrugated, or a diamond design. A surface with a corrugated

Figure 2-10   Matting
Courtesy of Salisbury Electrical Safety LLC

or diamond design is expected to be nonslip. However, if the surface of the mat is contaminated with a lubricant, the nonslip surface is ineffective.

Mats and matting may be constructed from either natural or synthetic rubber compound. Type I mats and matting may be made from any elastomer or elastomeric compound. Type I mats and matting must be properly vulcanized. Type II mats and matting have one of two special properties: Type II A material is resistant to ozone, and Type II B is flame resistant. A purchaser must indicate which type of mat or matting is required.

Mats and matting are available in standard voltage classes: Class 0, Class 1, Class 2, Class 3, and Class 4. Like other voltage-rated rubber products, the assigned voltage class is related to the maximum acceptable circuit voltage for which the product may be used.

The construction and initial testing of rubber mats and matting are addressed in ASTM D178, *Standard Specification for Rubber Insulating Matting*. The standard defines the required tests, how the tests are conducted, acceptable product criteria, and how the manufacturer responds to purchasers' inquiries.

ASTM subcommittee F18.25, which is a subcommittee of F18, develops and maintains ASTM D178.

ASTM D178 makes no provision for retesting of mats and matting. Purchasers must accept the responsibility to maintain the integrity of mats and matting after the product is delivered.

ASTM D178 accepts mats and matting in widths of 24, 30, 36, and 48 inches. Mats are provided in precut dimensions, and matting is provided as rolls in lengths as requested. Mats and matting are packaged as rolls or as flat bundles, but ASTM D178 requires packaging that will not distort the product.

The manufacturer must mark mats and matting to indicate the type and class. The marking must also indicate the name of the manufacturer and ASTM D178, provided the material complies with the requirements of this standard.

How Mats and Matting Provide Protection

Muscle tissue reacts to current flow through the body. Shock and electrocution are the result of current flow. Of course, voltage acts as the pressure that forces the current to flow. As the amount of current increases, the body reacts more violently. If the current flow can be limited to a value that has no deleterious effect, then neither shock nor electrocution is possible. The intended role of mats and matting is to reduce the amount of current flow should contact with an exposed energized conductor be made.

The objective is to prevent current flow within a worker's body. Additional insulating material may be inserted between the worker and an electrical conductor, or the additional insulating material may be inserted between the worker and earth. Voltage-rated gloves increase the impedance between the worker and an energized conductor. Mats and matting increase the impedance between the worker and the surface on which the worker stands.

If a worker touches an exposed energized conductor (and no other conductor or grounded surface) while standing on an appropriately rated mat, the amount of current flow through the worker's body will be small, and the worker will not be injured (see **Figure 2-11**).

For current to flow and for a worker to receive a shock, the worker must be a part of a completed circuit. If an insulating material is inserted between the worker and either of the components of a potential earth ground circuit, current cannot flow, and the worker cannot receive a shock.

Rubber gloves and blankets insert insulation between a worker and an energized component. The worker can be in contact with earth ground and still current cannot flow, because the gloves or blanket reduce the chance of contact with an energized component (see **Figure 2-12** and **Figure 2-13**).

Mats and matting are intended to insert insulation between a worker and earth ground. The worker can be in contact with an energized conductor, but current cannot

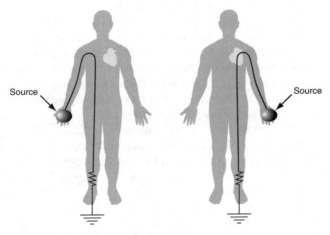

Figure 2-11   Current Flow through the Human Body

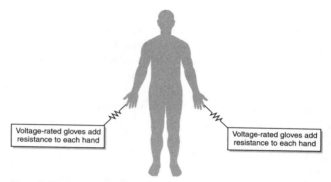

Figure 2-12   Rubber Gloves Add Resistance between the Worker and an Energized Part

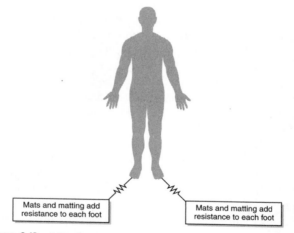

Figure 2-13   Mats and Matting Add Resistance between the Worker and Earth Ground

flow because the mat provides insulation between the worker and earth ground.

Mats and matting are less protective than voltage-rated gloves, because the mat or matting provides insulation only between the worker and the floor on which the worker stands. However, there are other grounded surfaces likely to be touched by the worker. Current flow from one hand to the other hand causes the same body reaction as current flow from one hand to a foot.

Degradation, Deterioration, and Damage

Mats are intended to provide a surface on which a worker may stand to perform work tasks. The surface of the mat or matting is exposed to damage from contaminants on the shoes of workers who walk on the surface. Although the voltage rating has been determined by testing when the mat or matting is installed, it is not easy to determine the amount of degradation after the mat or matting has been used for a period of time. Although it is possible to retest the mat or matting, retesting is somewhat impractical.

Initially, rubber mats and rubber matting are tested and assigned a voltage rating by the manufacturer. When the mat or matting is first installed, the voltage rating is dependable. However, after the mat or matting has been installed, walked on, cleaned, and damaged by foot and rolling traffic, the rating of the mat or matting has probably changed significantly. No realistic method exists that can verify that the initial rating has not deteriorated.

Some electronic equipment is easily damaged from a discharge of static voltage. Insulating mats and matting actually increase the opportunity for static buildup by increasing the impedance between the worker and the floor. Although the rubber mats and matting cushion a worker's feet with the floor, mats and matting should not be installed for use where sensitive electronic equipment is assembled or maintained.

Rubber mats and matting cushion the surface and relieve foot stress for workers who stand on the mat for long periods of time. However, the authors do not recommend these items for purchase to protect a worker from shock or electrocution. The opportunity is too great for the rubber to be damaged or contaminated quickly through normal use, and the degradation might not be noticeable. Thus the intended protection is no protection at all.

Purchase
Each purchase order for mats or matting must provide the following information:

- Type
- Class
- Thickness
- Width
- Length
- Color

Because an effective retest is not viable, mats or matting should not be included when shock-protective equipment is selected. When rubber mats or matting are installed near electrical equipment, only flame-resistant material should be used. The authors do not recommend these products for protection from electrical shock or electrocution.

Blankets

In normal use, blankets come into direct contact with the electrical conductor. Consequently, the prohibited approach boundary is penetrated each time a blanket is installed. Current consensus standards require each employer to determine necessary procedural and administrative actions when the prohibited approach boundary is penetrated. The circuit should be deenergized before the blanket(s) is (are) installed. The worker should ensure that the circuit is clear of any potential fault prior to reenergizing the circuit.

It is sometimes impractical to deenergize the circuit to install blankets for an emergency work task. In that instance, strict adherence to procedural requirements is critical. Blankets normally are wrapped around an electrical conductor in such a way that the worker is less likely to contact the conductor when executing the work task (see **Figure 2-14**). The blankets may be held in place by nonconductive cable ties, buttons, or clamp pins. Electricians frequently refer to the clamp pins as clothespins, because they have the general appearance of large clothespins. Blankets with hook-and-pile (Velcro®) fastening along the edges are also available.

Blankets are intended to provide temporary insulation on an electrical conductor and should not be left in place after the work task is complete.

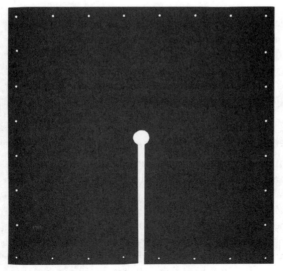

Figure 2-14   Blankets
Courtesy of Salisbury Electrical Safety LLC

Voltage Rating

Voltage rating of blankets follows the same voltage-class system used by other rubber insulating components. Blanket ratings are Class 0, Class 1, Class 2, Class 3, and Class 4. (See **Table 2-7** for blanket sizes and construction.)

Hazards

Because the intended function of insulating blankets is to increase the impedance or resistance between a worker and an energized conductor, blankets are effective in reducing a worker's exposure to shock and electrocution. However, insulating blankets are ineffective as protection from other electrical hazards such as arc flash or arc blast. Rubber blankets are held in place by clothespins, cable ties, or other means, and the device used to hold the blanket in

Table 2-7
Insulating Blankets

Style	Size *	Type / Material
Slotted (with or without eyelets)	22 in. × 22 in.	II / Synthetic rubber
	36 in. × 36 in.	II / Synthetic rubber
	46 in. × 46 in.	II / Synthetic rubber
Standard with eyelets	22 in. × 22 in.	II / Synthetic rubber
	27 in. × 36 in.	II / Synthetic rubber
	36 in. × 36 in.	I / Natural rubber
	36 in. × 36 in.	II / Synthetic rubber
	46 in. × 46 in.	II / Synthetic rubber
Standard without eyelets	18 in. × 36 in.	II / Synthetic rubber
	27 in. × 36 in.	II / Synthetic rubber
	36 in. × 36 in.	I / Natural rubber
	46 in. × 46 in.	II / Synthetic rubber

* Sizes are available with Class 2 and Class 4 ASTM ratings.
Source: Salisbury Electrical Safety LLC

place could become a missile should an arcing fault occur, thus introducing a new hazard.

An arcing fault is electrical current flowing in air between two conductors. Arcing faults may be the result of component failure. However, a worker making inadvertent contact with an energized conductor usually initiates an arcing fault. An adequately rated blanket reduces the chance of initiating a fault on the conductor. However, the blanket will have no effect in reducing a worker's exposure to the thermal and blast effects of an arcing fault.

An arcing fault also produces a significant pressure wave. Fasteners could become projectiles should an arcing fault occur. Placement of fasteners is an important consideration when positioning the fasteners on the temporary blanket. Buttons (**Figure 2-15**) and clothespins (**Figure 2-16**) are more likely to cause injury than cable ties.

Figure 2-15   Buttons
Courtesy of Salisbury Electrical Safety LLC

Figure 2-16   Clothespin
Courtesy of Salisbury Electrical Safety LLC

Selection and Use

Voltage-rated blankets are available in different sizes and colors. Users may wish to select specific colors and sizes for specific uses. The blankets also are available with slots that enable the blanket to be installed around a component such as an insulator or termination. Blankets are available with eyelets that accept buttons or cable ties. They also are available with Velcro®-type closures stitched to the blanket. The Velcro®-type closures enable rapid installation and removal. Note that although this type of fastener is effectively non-conductive, the fastener has no voltage rating.

Blankets are available as natural rubber (Type I) or elastomeric (Type II) construction. Type I blankets are subject to ozone checking, whereas Type II blankets resist ozone and ultraviolet degradation.

Voltage-rated blankets must have a voltage rating at least as high as the circuit voltage. For instance, Class 1 blankets (7500-volt rating) may be used on circuits where the voltage rating is 7.5 kV or less, but if the circuit voltage is greater than 7.5 kV, blankets with a higher rating must be used.

Selection of a blanket for a specific work task should include consideration of the blanket's style, size, and the fasteners that will be used to hold the blanket in the proper position to eliminate exposure to contact with the energized conductor. For instance, a blanket that is too large may leave openings at the edge of the blanket. Likewise, blankets without eyelets must rely on clothespins to secure the position. Blankets should be installed and secured to remain in place until the work task is complete, then removed and returned to storage.

Purchase

Purchase orders for blankets should contain a requirement for the product to comply with all requirements defined in

ASTM D1048, *Standard Specification for Rubber Insulating Blankets*.

The following information must be selected and provided on the purchase order:

- Size
- Voltage rating (class)
- Color
- Construction style (with or without eyelets)
- Type I or Type II
- Slotted (if required)

Storage

Voltage-rated blankets should be stored in canisters or storage containers specifically designed to protect the blanket from damage. Storage containers provide maximum protection. If the container is large enough, more than one blanket can be kept in the same container. However, the blankets should be rolled and inserted into the container separately. This method will permit one blanket to be removed without disturbing the remaining blankets. Some manufacturers provide roll-up style storage containers. Roll-up style containers require less space than canister containers, but they provide commensurately less protection.

After the blankets are inserted appropriately into the storage containers, the containers should be kept in a clean and dry location. Blankets should never be folded.

Maintenance Testing

Voltage-rated blankets must be tested routinely to ensure that the insulating integrity of the insulating material is maintained. Blankets must be tested as defined in ASTM F479, *Specification for In-Service Care of Insulating Blankets*, which describes the process used to determine if the blankets can still protect a user. The insulating integrity of blankets must be verified at intervals not exceeding 12 months. Employers must certify that the blankets have been tested within the previous 12 months. The certification may take the form of dated inspection records or ensuring that the date of the last test is marked on the edge of each blanket in non-conductive ink. The process for testing and retesting of voltage-rated blankets is discussed in 29 CFR 1910.137 (Electrical Protective Equipment) as well as ASTM F479.

Inspection

Rubber blankets normally are in intimate contact with exposed energized conductors; therefore, the insulating integrity of the blanket is extremely important. Rubber blankets should be inspected visually before each use. When insulating blankets are issued to a worker, the first action should be to remove the blanket from the storage container and make sure the last inspection occurred within the previous 12 months. If the test is out of date, the blanket should be marked out of service and returned to the tool crib for testing.

The worker should next conduct a visual inspection of the blanket surface. The complete physical inspection should assess damage from ozone, chemicals, cuts, abrasion, and penetration by a foreign object. Blankets are sometimes left in place for several days and, therefore, are exposed to damage from UV energy and chemicals such as petroleum-based lubricants. Normally, a close visual inspection will find soft spots or swollen areas in the surface of the blanket. This type of damage indicates exposure to petroleum-based lubricants or other chemicals. Rolling the blanket between the fingers while inspecting the surface area will find small penetrations or cuts. If damage is found during the visual inspection, the blanket should be tagged out of service and returned to the tool crib for retesting.

Cutting the damaged area from the blanket and discarding it is an acceptable method of repairing a damaged area. Although consensus standards indicate other repairs are acceptable, the only recommended method of repair is removing and discarding the damaged area. Note that if the remaining blanket is at least 22 inches by 22 inches, the blanket need not be destroyed.

Live-Line Tools (Hot Sticks)

Electricians have used the term "hot stick" for many years. The term suggests that the device is intended to contact an energized conductor. To most people (as well as in standard dictionaries), the word "hot" relates to temperature. When an electrician says that a conductor is "hot," he or she probably is talking about the fact that a conductor is energized relative to earth ground. However, the electrician could be talking about the thermal condition of a conductor that is overloaded. In another example, when thermographers perform analyses on electrical conductors and terminations, the term "hot" refers to the thermal condition of the conductor or termination. Use of the term "hot" to mean both "energized" and "high thermal temperature" can be very confusing.

The term "hot stick" then, is a misnomer, although it is the most commonly used term. OSHA regulations use the term "hot" to refer to a thermal condition instead of an energized condition. In the OSHA regulations, the term "hot stick" is used only in an appendix, and then only in an attempt to ensure complete understanding by the user. In 29 CFR 1910.269(j), OSHA uses the descriptive term "live-line" tools instead of "hot stick" to discuss tools constructed from fiberglass-reinforced plastic (FRP). In common usage, both "hot stick" and "live-line tools" refer to a family of tools constructed from FRP material that are used to perform specific actions on an energized conductor or compo-

Figure 2-17   "Shot-Gun" Live-Line Tool
Courtesy of Salisbury Electrical Safety LLC

nent (see **Figure 2-17** and **Figure 2-18**). To minimize confusion, this book uses the term live-line tools.

Hazards

Live-line tools enable the user to contact an exposed energized conductor or component and perform a specific function. Frequently, an action to hot-tap an existing conductor (make a service connection on an energized line) is necessary to add a new customer or unit. A properly assembled live-line tool enables a worker to complete the tap without encroaching on the restricted or prohibited approach boundary with hands or feet.

The recognized hazards associated with an energized electrical conductor are both shock and arc flash (see Chapter 3). The FRP material from which live-line tools are made is nonconductive and provides significant impedance between a worker and the energized component, as long as the surface of the tool is in good condition. When used properly, the live-line tool prevents exposure to electrical

Figure 2-18   Live-Line Tool Construction

shock. However, live-line tools that are contaminated with dirt, oil, or other conductive materials might expose a worker to an unrecognized hazard.

Should an arcing fault occur when the worker is using a live-line tool, the worker could be exposed to the thermal hazard associated with the arcing fault. The live-line tool provides no protection from this hazard. The chance of initiating an arcing fault by contacting an energized component with a live-line tool is remote. However, equipment or devices that may be attached to the end of a live-line tool can initiate an arcing fault (see **Figure 2-19**). The length of

Figure 2-19   Arcing Fault on an Overhead Line
Courtesy of D. Ray Crow

the live-line tool is the critical component of preventing thermal injury to an otherwise unprotected worker.

For many years, live-line tools were made from dry wood, such as maple. The tightly compressed fibers of such wood minimizes the tools' ability to absorb moisture. Some wooden live-line tools might still be in use. The surface of wooden live-line tools is more susceptible to contamination or moisture intrusion than the surface of FRP live-line tools, which makes the visual inspection very important. The authors recommend that employers replace all live-line tools made from wood with devices of FRP construction.

Live-line tools generally are constructed from foam-filled, hollow FRP, but they may be solid FRP construction without a center hole (i.e., a rod). The assembled live-line tool may be a combination of these two different constructions. For instance, an attachment is available that enables the worker to close and open a clamp at the top end of the live-line tool. This arrangement is commonly called a shot-gun assembly because the action resembles the cocking action associated with a shotgun.

ASTM F711, *Standard Specification for Fiberglass-Reinforced Plastic (FRP) Rod and Tube Used in Live-Line Tools*, defines requirements associated with the construction and testing of live-line tools. This standard is produced and maintained by the subcommittee F18.35, and it defines the characteristics of the rod- and foam-filled tube. However, the standard does not define attachments and fittings. The rod- and foam-filled tube provides the necessary electrical insulation. The electrical characteristics of the attachments have no bearing on the insulating characteristic of the overall assembly.

Selection and Use

Customers expect that utilities will keep electrical energy available for their use in homes, businesses, commercial

endeavors, and industrial locations. Utilities and manufacturers respond to that demand by developing equipment,
tools, and work processes that enable maximum system
availability for customers. Customers also expect that utilities will provide the electricity at minimum costs. Utilities
respond by developing equipment that minimizes the cost
of the installation to the consumer while maintaining the
integrity of the service. For instance, fuse clips become disconnecting devices. Devices that permit an appropriately
assembled live-line tool to operate the disconnecting device
replace ground-operated switching devices. Frequently, the
means to disconnect the utility will be located at the top of
a pole (see **Figure 2-20**). A live-line tool enables a worker

Figure 2-20   Live-Line Tool Disconnect Switch

to operate the disconnecting means from the ground or from a bucket truck.

Disconnecting means that are operated by a live-line tool cost much less than gang-operated switches (see color insert, **Figure CP-2**). (Gang-operated means that all phases are connected together mechanically so that all phases break and make together.) Because these devices are very economical, they are sometimes fitted into a cabinet and act as a disconnecting device for pad-mounted transformers. Commercial facilities and small manufacturing facilities frequently use pad-mounted transformers with live-line tool-operated disconnecting means.

Utility workers use live-line tools to perform most functions on energized distribution and transmission lines that serve end-user facilities. Workers use a live-line tool when replacing fuses and operating disconnecting means. When a transmission or distribution line is deenergized for major maintenance, repair, or installation, workers install grounding devices with an appropriately assembled live-line tool. Grounding devices are necessary when performing line maintenance or repair. Workers consider the transmission or distribution line to be energized until safety grounds are installed.

Generally, larger manufacturing facilities require immediate access to disconnecting means. These facilities normally install gang-operated switches, enclosed fusible switches, or circuit breakers in an enclosure. This installation enables workers to operate the handle of the equipment quickly to deenergize the electrical service. However, some industrial facilities employ pad-mounted transformers with disconnecting devices that are operated using live-line tools (see **Figure 2-21**). Thus, industrial workers must follow the same work processes and practices as utility workers.

Accessories are available that can equip the live-line tool to perform most functions. For example, the live-line

Figure 2-21   Pad-Mounted Transformer

tool can be utilized as a tool for rescue workers who might be in contact with an energized conductor, or the tool can be equipped with grounding equipment to discharge static voltage (see **Figure 2-22**).

When using a live-line tool, workers must observe the minimum approach distance. Table R-6 in OSHA 29

Figure 2-22   Static Discharge Stick
Courtesy of Salisbury Electrical Safety LLC

CFR 1910.269 defines minimum approach distances for workers. Workers must not approach an energized circuit more closely than the minimum approach distance specified in Table R-6. However, the authors recommend using the minimum unprotected approach distance provided in Table 2-1 in this book. The distances in this table are taken from the restricted and prohibited approach distances in NFPA 70E, Table 130.2(C). The restricted approach distance provided in Table 2-1 meets or exceeds the phase-to-ground minimum approach distances defined in Table R-6 in OSHA 29 CFR 1910.269 for clear, live-line tool distances.

Workers must execute a hazard/risk analysis as defined in NFPA 70E-2004 before performing a work task associated with the use of a live-line tool. If the analysis suggests exposure to a potential arcing fault, the worker must be protected from the thermal hazard. Workers must not wear clothing that could increase the extent of the injury. Clothing made from acetate, nylon, polyester, and rayon is specifically forbidden.

Long live-line tools sometimes are difficult to manage. Workers must be particularly aware of how their bodies are positioned. When using a live-line tool, workers must be very conscious of the length of the tool so that they do not create the additional hazard of inadvertently contacting energized equipment.

Purchase
A purchase order should contain a requirement for the live-line tool to comply with the requirements defined in ASTM F711. The purchase order must define many different characteristics. The purchase order must define the length and diameter, as well as all accessories that are required. Without accessories, the live-line tool is simply a foam-filled tube. The purchase order does not need to define voltage

rating. ASTM F711 defines the test voltage for all live-line tools.

When new, all FRP live-line tools must be tested at 100 kV per foot of length for a period of five minutes without detectable heating. Wooden live-line tools must be tested at 75 kV per foot of length without detectable heating. Note that although wood live-line tools are discussed in consensus standards, the authors do not recommend their purchase.

A storage container should be purchased for each live-line tool at the time the purchase order is placed. Storage containers are available in several different constructions; the purchase order must define the required construction type.

Storage

Live-line tools should be stored in a container fabricated for that purpose. The storage container should provide protection from potential contamination and damage. When the live-line tool will be assigned to a service or line truck, the storage container should be of rigid construction and mounted on the vehicle. If the live-line tool is to be kept in a tool crib or similar location, vinyl or canvas containers might provide the necessary protection.

Maintenance Testing

When they are new, live-line tools must be tested as defined in ASTM F711 (i.e., 100 kV per foot of length for FRP construction). However, after the device has been in use for a period of time, the consensus standard that addresses testing shifts to IEEE Standard 978, *Guide for In-Service Maintenance and Electrical Testing of Live-Line Tools*.

Installations that use live-line tools for employee protection must remove the live-line tools from service at intervals no greater that two years and verify their insulating quality. FRP live-line tools must be tested at 75 kV per foot of length for one minute. No discernable heating should be detected. If the device is of FRP construction, the test must verify that the insulating quality remains, even in wet conditions.

Wooden live-line tools must be tested at 50 kV per foot of length with no discernable heating. Wooden live-line tools should not be used in wet or high-humidity conditions.

Inspection

Consensus and regulatory standards require live-line tools to be visually inspected each day before use. The authors recommend that the device be inspected completely before each use. The inspection should check specifically for damage and contamination. If any damage is found, the live-line tool should be tagged accordingly and removed from service. If the live-line tool can be repaired to its original condition, the device may be subjected to the maintenance test defined in IEEE Standard 978 and returned to service. However, an employer probably will find that replacing the damaged live-line tool with a new one is the most economical option.

The live-line tool should be cleaned and waxed before each use. (If the live-line tool is used more than once in a single day, one cleaning before the first use is sufficient.) Although the insulating integrity of the live-line tool is defined and the tool is visually inspected before use, the authors strongly recommend that the worker wear voltage-rated rubber gloves that are adequately rated for the circuit voltage.

Footwear

Workers are exposed to shock and electrocution when electrical current flows through the body. The reaction of the worker's body changes as the amount of current increases.

Creating an electrically safe work condition is the best way to control current flow through the worker's body. However, in some instances, removing the source of electricity is not a viable option. An alternative way to control current flow through the worker's body is to select and wear adequately rated PPE.

For current to flow through a worker's body, the worker must contact an exposed energized conductor and another conductor, such as earth ground. If the worker wears PPE on his or her hands that is rated for the voltage, the potential for the hands being the point of entry for electricity is reduced or eliminated. If the worker wears footwear that is adequately rated, foot contact with earth ground is reduced or eliminated. However, neither PPE for the hands or the feet prevent other body parts from contacting an energized conductor or earth ground. It is important to recognize that protecting the hands and feet does not protect arms, legs, and other body parts.

Dielectric footwear (see **Figure 2-23**) must be constructed as defined in the standard ASTM F2413. The technical detail of dielectric footwear construction is under the jurisdiction of the ASTM F13.30 subcommittee. This committee writes and maintains standard ASTM F2413, *Standard Specification for Performance Requirements for Foot Protection*, which defines the protective nature of the footwear for hazards other than electrical shock. Tests to determine the electrical protective nature of dielectric footwear are defined in ASTM F1116, *Standard Test Method for Determining Dielectric Strength of Dielectric Footwear*. These national consensus standards provide assurance that the footwear meets normal expectations for protecting a worker's feet from both physical and electrical shock hazard.

Dielectric boots and shoes could be worn as ordinary shoes as long as the boots or shoes fit the worker's feet

Figure 2-23   Dielectric Footwear
Courtesy of Salisbury Electrical Safety LLC

properly. However, overshoes and boots are intended to be worn in addition to (on top of) ordinary work shoes.

Impact on Exposure to Electrical Hazards

Dielectric footwear provides additional impedance to help isolate a worker from contact with earth. Manufacturers test dielectric footwear to a predetermined voltage level (20 kV or 14 kV). The footwear must be marked to indicate the voltage rating of the overshoes or boots.

Consensus and regulatory standards do not require footwear to have a voltage rating. However, their use is certainly permitted. Dielectric footwear was developed and sold on the market for many years when the only recognized personal electrical hazard was shock and electrocution. When a worker is located at the top of a utility pole, the worker is likely to be in contact with the pole with both feet. Should he or she contact an open energized conductor with one hand, the dielectric footwear will prevent shock injury or electrocution. Footwear was seen as one option to reduce the risk of current flow through a worker's body, and manufacturers developed dielectric footwear for use by these workers. However, no requirement for dielectric rated footwear exists, and few workers use it.

Because manufacturers designed dielectric footwear for use by workers climbing utility poles, the footwear is constructed so that the heel will grip the rungs installed on the pole. The sole of the footwear is solid and provides support for the worker's feet when he or she is wearing climbers.

In some instances, workers who are excavating a hole in the earth or in a concrete slab could be exposed to contact with an existing energized electrical conductor. Should such contact occur, dielectric footwear could eliminate exposure to step potential.

Purchase
Purchase orders for dielectric footwear should contain a requirement for the product to comply with the requirements of ASTM F1117. The purchase order also must provide the following information:

- Description—boots (including height) or overshoes
- Size

Since an effective retest is not viable, dielectric footwear should not be included when shock protective equipment is selected. The authors do not recommend these products as primary protection from electrical shock or electrocution. Dielectric footwear should always be considered as secondary protection, at best.

References

ASTM D120, *Standard Specification for Rubber Insulating Gloves*. Conshohocken, PA: American Society of Testing and Materials, 2002.
ASTM D178, *Standard Specification for Rubber Insulating Matting*. Conshohocken, PA: American Society of Testing and Materials, 2002.

Figure CP-1 Worker Wearing Sleeves Correctly
Source: © David R. Frazier Photolibrary, Inc./Alamy Images

Figure CP-2 Operating a Disconnect Switch
Source: © Matthew Collingwood/ShutterStock, Inc.

Figure CP-3 FR-Rated Switching Hood
Source: © 2007, VF Imagewear, Inc. Used with permission.

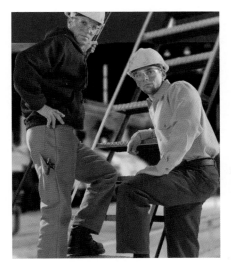

Figure CP-4 FR Work Clothes
Courtesy of Workrite Uniform Company

Figure CP-5 FR Clothing for Women
Courtesy of Workrite Uniform Company

Figure CP-6   Multimeter

Figure CP-7   Voltmeter
Courtesy of Tegam

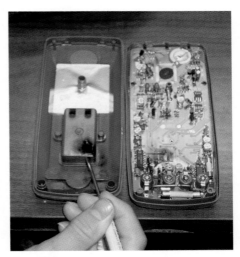

Figure CP-8   Failed Voltmeter

Figure CP-9   Non-Contact Meter
Courtesy of Fluke Corporation. Reproduced with permission.

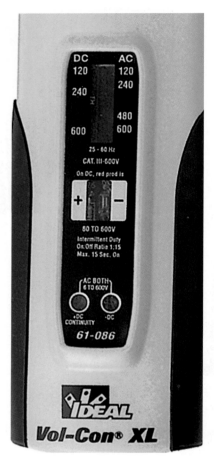

Figure CP-10 Label Showing Duty Cycle
Courtesy of Ideal Industries, Inc.

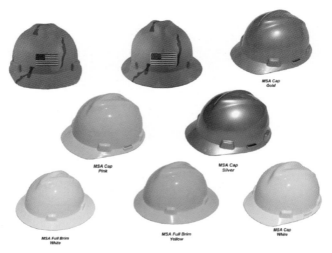

Figure CP-11 Protective Helmets (Hard Hats)
Courtesy of Tasco-safety.com

Figure CP-12 Arc-Rated Face Shield
Courtesy of Salisbury Electrical Safety LLC

Figure CP-13 Overhead Line Work
Courtesy of Win Henderson/FEMA

Figure CP-14 Worker Installing a Ground Cluster
Courtesy of Salisbury Electrical Safety LLC

ASTM D1048, *Standard Specification for Rubber Insulating Blankets.* Conshohocken, PA: American Society of Testing and Materials, 2002.

ASTM D1051, *Standard Specification for Rubber Insulating Sleeves.* Conshohocken, PA: American Society of Testing and Materials, 2002.

ASTM F479, *Specification for In-Service Care of Insulating Blankets.* Conshohocken, PA: American Society of Testing and Materials, 2002.

ASTM F496, *Specification for In-Service Care of Insulating Gloves and Sleeves.* Conshohocken, PA: American Society of Testing and Materials, 2002.

ASTM F711, *Standard Specification for Fiberglass-Reinforced Plastic (FRP) Rod and Tube Used in Live-Line Tools.* Conshohocken, PA: American Society of Testing and Materials, 2002.

ASTM F1116, *Standard Test Method for Determining Dielectric Strength of Dielectric Footwear.* Conshohocken, PA: American Society of Testing and Materials, 2005.

ASTM F2413, *Standard Specification for Performance Requirements for Foot Protection.* Conshohocken, PA: American Society of Testing and Materials, 2005.

IEEE Standard 978, *Guide for In-Service Maintenance and Electrical Testing of Live-Line Tools.* Piscataway, NJ: Institute of Electrical and Electronics Engineers, 1984.

NFPA 70E, *Standard for Electrical Safety Requirements for Employee Workplaces.* Quincy, MA: National Fire Protection Association, 2004.

U.S. Department of Labor. Occupational Safety and Health Administration. OSHA Regulations 29 CFR 1910.137, "electrical protective equipment." Washington, DC.

U.S. Department of Labor. Occupational Safety and Health Administration. OSHA Regulations 29 CFR 1910.269, "electric power generation, transmission, and distribution." Washington, DC.

Protection from Arc Flash

▣ Introduction to FR-Rated Protection

When the use of electricity was just beginning to expand, no consensus standard existed for electrical installations. Insurance records indicate that fires and electrocutions were common. Economic losses associated with these incidents were costly.

Several agencies came together to study the cause of these losses and determined that the primary reason for the incidents was the lack of a common method to install electrical conductors and facilities. Consensus standards and codes grew from a common objective to eliminate or reduce the number of fires and electrocutions from the use of electricity as an energy source. Training programs, equipment construction, and other processes emerged from the recognition that developing equipment and installation standards could reduce or eliminate the unsafe conditions.

Arcing Faults

As equipment was maintained and repaired, arcing faults occurred. Electricians generally recognized that when an arcing fault occurred, these would be the normal results:

- Parts and pieces would fly from the enclosure.
- Bus and components would melt.
- Weld splatter would be deposited on the worker's glasses and clothes.
- The worker frequently would be burned or injured.

Even with this experience and knowledge, the arc associated with the fault was not recognized as presenting a hazard to the worker. A burn injury was recorded as a burn, with no association to electrical energy.

In the mid-1980s, several organizations began to understand that results of an arcing fault were unacceptable. Professional white papers and other publications began to discuss arc flash. That recognition of unacceptable conditions had two components: the arc-generated temperatures that caused burns to the worker and damage to the equipment caused by high temperatures. Members of the NFPA 70E Technical Committee recognized that the degree of injury increased greatly in instances where the worker's clothing ignited or melted into the injured worker's skin.

NFPA 70E Recognizes Arc Flash

With the publication of the 1995 edition of NFPA 70E, the committee formally recognized the thermal hazard associated with an arcing fault as an electrical hazard and recognized that workers needed to be protected from the thermal hazard. Although equipment that would protect a worker from the thermal exposure of an arcing fault had not been developed, protective equipment for other types of thermal exposure were in common use. For example, drivers of racecars and firefighters routinely wore flame-resistant clothing for protection. Some employers required workers in areas of high risk of a fuel-based fire to wear flame-resistant clothing at all times.

FR Clothing

Members of the NFPA 70E Technical Committee agreed that eliminating the risk of clothing ignition or melting would prevent many injuries and reduce the severity of the remainder. The committee also agreed that the flame-resistant clothing available at the time would reduce the severity of burn injuries. At the very least, wearing flame-resistant clothing would eliminate the risk associated with clothing ignition.

At the same time, the Occupational Safety and Health Administration was in the process of drafting and publishing 29 CFR 1910.269 in Subpart R. This revision prohibited clothing that could increase the degree of injury from exposure to an electrical arc. The same section of the OSHA rules stated that clothing made from polyester, acetate, nylon, or rayon were prohibited from use by workers who might be exposed to an arc flash event. The OSHA rules permitted (and still do allow) clothing made from these materials to be worn, provided the employer can demonstrate that the clothing is treated to withstand the conditions or to eliminate exposure to the potential hazard.

After the promulgation of NFPA 70E-1995 and the revision to 29 CFR 1910.269, ASTM developed standards that defined tests necessary to determine if clothing met protection standards. ASTM issued ASTM F1506, *Standard Performance Specification for Flame-Resistant Textile Materials for Wearing Apparel for Use by Electrical Workers Exposed to Momentary Electric Arc and Related Thermal Hazards*, and clothing manufacturers began to construct clothing that complied with the standard. The manufacturers designated various levels of protection, and clothing began to be available with the protective characteristics printed on the label or other places on the garment. Clothing that meets the requirements of the current edition of the ASTM standard is considered flame-resistant (FR) clothing. Fabric or other material that is flame resistant resists ignition. The flame-resistant characteristic remains with the clothing for the life of the garment. (For a list of standards that regulate arc flash protective equipment, see **Table 3-1**.)

Table 3-1
Standards for Arc Flash Protection

Subject	Number and Title
Apparel	ASTM D6413, *Standard Test Method for Flame-Resistance of Textiles*
	ASTM F1449, *Standard Guide for Care and Maintenance of Flame-Resistant Clothing*
	ASTM F1506, *Standard Performance Specification for Flame-Resistant Textile Materials for Wearing Apparel for Use by Electrical Workers Exposed to Momentary Electric Arc and Related Thermal Hazards*
	ASTM F1958, *Standard Test Method for Determining the Ignitability of Non-Flame-Resistance Materials for Clothing by Electric Arc Exposure Method Using Mannequins*
	ASTM F1959, *Standard Test Method for Determining the Arc Thermal Performance Value of Materials for Clothing*
Gloves and sleeves	ASTM D120, *Standard Specification for Rubber Insulating Gloves*
	ASTM D1051, *Standard Specification for Rubber Insulating Sleeves*
	ASTM F496, *Standard Specification for In-Service Care of Insulating Gloves and Sleeves*
	ASTM F696, *Standard Specification for Leather Protectors for Rubber Insulating Gloves and Mittens*
Rainwear	ASTM F1891, *Standard Specification for Arc- and Flame-Resistant Rainwear*
Face protective products	ASTM F2178, *Standard Test Method for Determining the Arc Rating of Face Protective Products*

ASTM–American Society for Testing and Materials

Consensus requirements for FR clothing do not limit the materials of construction to particular fabrics or materials. In addition to materials that are inherently flame resistant, cotton and other fabrics can be treated with an agent to temporarily change the ignition characteristic of the fabric. Generally, the agent is a chemical that imparts a characteristic that retards the spread of flame after ignition. Although the product may pass the generally accepted vertical flame test, the characteristic is temporary. Each time the garment is laundered, the product's ability to resist ignition is reduced. The garment becomes less protective with each laundering until the protective characteristic essentially disappears. Treating a meltable fabric with a flame-retardant chemical may cause the fabric to resist flame spread, but such treatment does not change the inherent melting property.

Incident Energy Rating

Flame-resistant protective equipment is rated on the basis of incident energy. Incident energy is the thermal energy that might impinge on the exposed (or at-risk) surface. The incident energy is defined in terms of the total calories of thermal energy that might be received on a one-centimeter square section of the surface. As the incident energy rating of clothing increases, the number of calories per square centimeter increases, and the protective nature of the garment also increases. For instance, a garment rated at 15 calories per square centimeter is more protective than a garment rated at 8 calories per square centimeter.

An incident energy rating of FR clothing that is determined by testing in accordance with ASTM F1959, *Standard Test Method for Determining the Arc Rating of Materials for Clothing*, means that the clothing will prevent a second-degree burn 50 percent of the time at the assigned rating. To ensure avoiding a second-degree burn, the worker must wear clothing with a higher incident-energy rating. Without saying so directly, current consensus standards suggest that the risk associated with the current rating system is acceptable. A second-degree burn generally is described as a curable burn; the skin tissue will regenerate without scarring. To achieve greater assurance of avoiding all burns, consensus requirements for protective equipment must be exceeded.

FR clothing fails in two modes, and the assigned arc rating is based on the failure modes. In the first failure mode, the incident energy exceeds the thermal insulating ability of the garment. In the second failure mode, the garment chars and breaks open (breakthrough) to expose the surface under the garment. The test described in ASTM F1959 determines the thermal exposure that produces either result and identifies it as arc rating. Arc thermal performance value (ATPV) relates to the incident energy level that will result in conducting thermal energy of 1.2 calories per square inch or greater. Energy breakthrough (E_{BT}) indicates the incident energy level at which breakthrough occurs, that is, at which the charred FR material breaks open to expose the surface below. Both ATPV and E_{BT} are expressed in calories per square inch. Protective clothing that has an ATPV or E_{BT} rating may be used at incident energy exposure levels equal to or less than the ATPV or E_{BT} rating of the clothing. The clothing label will indicate the arc rating as ATPV or E_{BT}.

Label Requirements

ASTM F1506 requires the clothing manufacturer to provide certain information on the label in each garment (see **Figure 3-1**). The information is intended to provide a worker with sufficient information to properly select and use FR protection. The label must contain at least the following information:

- A tracking code
- Indication that the garment conforms to ASTM F1506
- Manufacturer's name

- Size and associated standard label information
- Care instructions and fiber content
- Arc rating, either ATPV or E_{BT}

A worker must determine that each item of FR clothing is suitable for the expected exposure by comparing arc rating on the clothing label with the incident energy exposure level identified in the hazard/risk analysis. When a worker performs the hazard/risk analysis required by NFPA 70E, he or she will know the incident energy level of potential exposure. By reading the label in the FR clothing, the worker knows what level of thermal protection can be expected from the FR clothing. If the arc rating of the equipment listed on the clothing label exceeds the expected exposure, the FR clothing is acceptable for that work task.

Thermal Barrier

Electrical shock protection in the form of rated rubber products reduce the risk of electrical shock or electrocution by reducing the amount of current that might flow in a

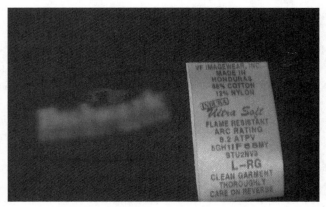

Figure 3-1   Clothing Label

worker's body. The rubber products introduce an additional barrier that resists the flow of electrical current. Likewise, FR clothing reduces the risk of thermal burn by reducing the amount of thermal energy that might flow onto a worker's body. The FR clothing introduces an additional barrier, which resists the flow of thermal energy.

To avoid a thermal burn injury, the additional thermal barrier must offer sufficient resistance to the flow of thermal energy to avoid the possibility of burn. Air is a relatively good thermal insulator. If a worker wears two sets of FR clothing, a layer of air is trapped between the layers of FR clothing. Therefore, the available thermal protection is the overall sum of the protection afforded by each layer of FR clothing, plus the protection provided by the air layer trapped between the clothing layers. The protection provided by layers is greater than the arithmetic sum of each layer of FR clothing. In some instances, manufacturers have tested various layer configurations and have assigned a rating for specific layering systems. However, consensus has not been achieved on the overall subject of layering.

Wearing flame-resistant clothing effectively prevents being burned from clothing igniting. If a worker wears FR clothing that is underrated, an injury could occur from the intense temperature of the arc, but not from burning clothing. Underrated clothing might conduct thermal energy to the person's skin, resulting in an injury. However, wearing FR clothing ensures that the clothing will not ignite and the injury's severity will not be increased from burning clothing, even if the clothing is underrated.

Selection and Use

The most severe injuries from exposure to an electrical arc are the result of flammable or meltable clothing igniting. In general, the duration of an electrical arc is limited by the clearing time of the overcurrent device. Consensus standards and codes guide the selection of overcurrent devices. Overcurrent devices function by monitoring the amount of current flowing in the circuit; when the amount of current exceeds the rating of the overcurrent device, the circuit is opened, and the source of energy is removed. Contemporary codes and standards permit overcurrent devices to be large to reduce the chance of nuisance trips. Architects and engineers generally specify circuit breakers and fuses that will reduce the risk of fire in the building and protect the equipment from destruction should a fault occur. Fire was one of the first hazards associated with the use of electrical energy and remains a primary concern.

Electrical equipment usually is constructed from metal components such as copper or aluminum conductors, electrical insulating components, and steel structures or enclosures. The metal components of electrical equipment remain solid until the temperature is elevated to the melting point. A slightly higher temperature results in the liquid metal beginning to evaporate by boiling, becoming metal vapor. The change of state of the metal components begins when the temperature is in the range of 1800°F. Within a few hundred degrees, the metal components evaporate and become metal vapor.

Human tissue begins to be destroyed when the temperature is elevated to about 150°F and held at that temperature for one second. Cells are destroyed in one-tenth of one second when the temperature of the tissue is elevated to about 200°F. To avoid a burn injury, then, both the temperature and the duration of the exposure must be limited (see **Table 3-2**).

Some clothing ignites if raised to a temperature of about 700°F while other clothing ignites at a temperature of 1400°F. Several mitigating factors are associated with clothing ignition temperature. Clothing may be ignited from heat energy transmitted to the clothing surface by a

Table 3-2
Effect of Temperature on Human Skin

Skin Temperature	Time to Reach Temperature	Damage Caused
110°F	6.0 hours	Cell breakdown begins
158°F	1.0 second	Total cell destruction
176°F	0.1 second	Second-degree burn
200°F	0.1 second	Third-degree burn

Source: Jones, Ray A. and Jane G. Jones, *Electrical Safety in the Workplace*, Sudbury, MA: Jones and Bartlett Publishers, 2000.

combination of radiation and convection. However, ignition also can result from conduction from metal droplets expelled from the arc.

Neither tissue damage nor clothing ignition occurs instantaneously. The duration of the exposure affects the resulting cell damage or clothing ignition. If an exposure is limited to a time period less than necessary for cell damage or clothing ignition, injury does not occur and clothing does not ignite or burn. If the clothing does not burn after the arc is removed, damage is limited to direct exposure to the arc and to heat conducted through the clothing covering the tissue.

Limiting Fault-Current Time

Overcurrent devices remove the source of energy from the arc in a predetermined period of time. Overcurrent devices do not clear the arcing fault instantaneously. When a fuse element begins to melt, current continues to flow until the opening in the melting element is large enough to break the circuit. Fuses contain a substance similar to sand, which flows into the opening in the melting element and quenches the arc. If the fuse is a current-limiting fuse and the fault current is large, the element begins to melt in one quarter of one cycle, or about 4.5 milliseconds, and clears the fault in less than two cycles, or 34 milliseconds. If the current in the arcing fault is below the current-limiting range of the fuse, however, the arcing fault might continue for several seconds.

Current-limiting circuit breakers take slightly longer than fuses to clear the fault, because a circuit breaker requires a component to physically move from one position to another. However, if the device is designated as current limiting, the duration of the arcing fault will be short unless the fault current is less than the current-limiting rating of the circuit breaker.

If uncontrolled, the temperature in an electrical arc could reach 35,000°F. For comparison purposes, the surface of the sun (as defined by astronomers) is generally accepted to be 9000°F. Overcurrent devices control the duration of an arcing fault. If the overcurrent device is current limiting, the duration is very short, which limits the temperature in the plasma. If the overcurrent device is not current limiting, the fault current is likely to continue for one half second to several seconds. As the arc current is permitted to continue, the temperature in the arc plasma continues to climb higher. Although the temperature is unlikely to reach 35,000°F, it is common for an arc temperature to reach 16,000° to 20,000°F.

To avoid injury, a worker must select protective clothing that has the following characteristics:

- The clothing must not ignite and continue to burn after the arcing fault has been removed.
- The clothing must not melt onto or into the worker's skin as a result of being exposed to the arc.
- The clothing must provide sufficient thermal insulation to prevent the worker's skin tissue from being heated to destruction.

The first two characteristics are much more important than the third. Regardless of the duration of the exposure, if the clothing does not ignite and does not melt onto the worker's skin, the duration of the resulting exposure will be limited to the time required for the overcurrent device to clear the fault. Workers who are or might be exposed to an arcing fault must wear clothing that will not melt and clothing that will not ignite and continue to burn after the overcurrent device clears the fault.

For an illustration that shows the relationship of clothing and ignition, see **Table 3-3**.

Transmitted Energy

The worker must determine the amount of thermal energy transmitted from the arc to select flame-resistant clothing that provides adequate thermal protection. The current state of knowledge can determine only an estimated amount of transmitted energy. Many factors influence the amount of en-

Table 3-3
Ignition Energies of Cotton Fabrics

Fabric Description			
Weight (oz/yd^2)	Weave	Type of Material	Ignition Threshold (cal/cm^2)
5.2	Twill	Shirt	4.6
6.2	Fleece	Shirt	6.4
6.9	Twill	Shirt	5.3
8.0	Twill	Shirt or Pants	6.1
8.3	Sateen	Shirt or Pants	11.6
11.9	Duck	Shirt or Pants	11.3
12.8	Denim	Jeans	15.5
13.3	Denim	Jeans	15.9

Source: IEEE Paper No. PCIC-97-35

ergy received by the worker's clothing or skin, including the following:

- The capacity of the circuit to supply energy (called available fault current)
- The clearing time of the overcurrent device
- Intervening barriers or the position of the worker's body
- The reflectivity of the enclosure (influenced by color) and the reflectivity of the worker's clothing or skin
- The length of the arc
- The distance of the arc from the worker

Change in any one of these factors changes the amount of energy received by the worker's clothing or skin.

Energy received by a worker's clothing or skin is called incident energy, or the amount of energy that is incident on a surface. Several different methods of estimating incident energy have been developed and are available in published literature. Annex D of NFPA 70E-2004 discusses several methods that can be used to calculate incident energy. Other methods also exist. An employer should conduct an arc flash analysis for electrical equipment on a facility and install a label on the front of each piece of equipment indicating incident energy available in the equipment. Workers can then determine available incident energy from the information on the label and select protective equipment that is rated at least as great as the amount of incident energy available.

Calculating Incident Energy

NFPA 70E-2004 requires a hazard/risk analysis (see Chapter 1). One aspect of a hazard/risk analysis requires workers to determine whether risk of exposure to a thermal hazard exists and, if such a hazard exists, the degree of the thermal hazard. A worker must determine the flash protection boundary to understand if he or she might be exposed to an arc flash. If any part of a worker's body will be within the flash protection boundary, PPE is necessary.

The Flash Hazard Analysis

The worker must determine the degree of the hazard by determining the amount of incident energy that will impinge on his or her clothing or skin. Current consensus standards require that a label be installed on specific electrical equipment that contains an arc flash hazard. Although no requirement exists for the flash protection boundary to be on the label, labels normally identify both the flash protection boundary and incident energy. The information that is listed on the label must be determined by estimating the thermal energy that would be released in an arcing fault at the location of the work task.

The flash hazard analysis process consists of three steps. Unless all steps are followed, the flash hazard analysis has not been performed. The steps are as follows:

1. Determine the flash protection boundary for the task.
2. Determine the incident energy for the task.
3. Select PPE that reflects the results of the first two steps.

If the worker performs the hazard/risk analysis correctly, he or she may realize that flash hazard protection is unnecessary. The worker may find that the work must be done with the equipment in an electrically safe work condition (with the equipment deenergized). The worker also might identify a work process that avoids the exposure and does not penetrate the flash protection boundary.

Lee Equations

The Lee equations were developed to predict the flash protection boundary. These equations are illustrated in Appendix D, Section D-3 in NFPA 70E-2004. In the original form, these equations predict the distance from an arcing fault that produces a curable burn. In Lee's vernacular, the term curable burn can be translated to second-degree burn. A curable or second-degree burn results from the skin receiving 1.2 cal/cm^2. The Lee equations determine the flash protection boundary.

Doughty Equations

The Doughty equations were developed to predict the amount of incident energy that might be received by a person working at a distance of 18 inches from an arcing fault. These equations were developed by R. L. Doughty, T. E. Neal, and H. L. Floyd and presented in an IEEE paper at the Petroleum and Chemical Industry Committee conference in 1998. These equations are illustrated in Appendix D, Section D-6 in NFPA 70E-2004. The data used to develop the equations assumed that the thermal energy transfer is essentially by radiation. These equations predict incident energy when the arcing fault is contained in an empty box and when the arcing fault is located in open air.

IEEE 1584

An IEEE working group gathered additional data and performed regression analyses. The working group assumed that the thermal energy is transferred essentially by radiation. These equations predict the incident energy at a distance of 18 inches for utilization equipment and at 36 inches for distribution equipment. The equations contained in IEEE Standard 1584, *Guide for Performing Arc-Flash Hazard Calculations*, are basically an extension of the Doughty equations.

Commercial Software

Several developers of commercial software utilize equations to calculate incident energy. The software is used to perform circuit analyses of distribution and utilization circuits.

Generally, the software utilizes equations similar to those offered in IEEE Standard 1584.

ARCPRO

Ontario Hydro Technologies in Toronto developed this commercially available software. The current copyright holder is Kinectrics, in Toronto, Canada. The software predicts incident energy based on circuit parameters that are similar to the parameters contained in both the Doughty equations and the IEEE 1584 equations. ARCPRO includes factors to account for energy transfer by both convection and radiation, in addition to variables such as properties of the gasses in the arc plasma. However, the equations used in the software are confidential and available from Kinectrics only.

Shareware

Some shareware programs are available on the Internet. The validity of the equations used in the shareware programs has not been verified. Users are cautioned not to use shareware programs to calculate incident energy.

The Flash Protection Boundary

One primary purpose of an arc flash analysis is to determine the flash protection boundary. The label on the equipment should identify the flash protection boundary, usually either in meters or feet and inches. Workers should determine the flash protection boundary from the label on the equipment. The flash protection boundary is a spherical shape, measured from the potential arc to any part of a worker's body (see **Figure 3-2**). When any part of a worker's body is close to the source of a potential arc, a thermal burn injury is possible and the worker's body must be protected from a possible burn.

Some people have used the misnomer arc flash boundary instead of the correct term. The term arc flash boundary should never be used. The reason is twofold: (1) The

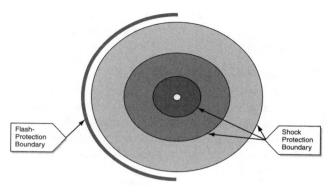

Figure 3-2   Flash Protection Boundary

boundary of the arc is unimportant, because the real issue is at what distance an injury would occur. (2) The focus should be on how far the worker should remain away from the point of the arc flash to remain uninjured, not on how far the arc could reach. The correct term, as defined in NFPA 70E, is flash protection boundary.

Point of Exposure Distance

Current methods of determining incident energy arbitrarily assign a distance of 18 or 36 inches as the point of exposure. These dimensions were selected because a worker's chest area is likely to be at approximately that distance from a potential arcing fault. A worker's hands (and face perhaps) may well be closer than 18 inches. The worker's hands (and face), then, are exposed to a greater thermal hazard and should have greater protection. Although some manufacturers provide gloves made from the same materials as clothing, no formal rating system for arc flash has been established. However, experience suggests that, in most cases, heavy-duty leather gloves (such as those worn with voltage-rated gloves) effectively protect a worker's hands and arms. The only option for the face is to wear a face shield or a hood. This determination should be made by the flash hazard analysis.

Other Exposure Considerations

When FR clothing is necessary for a work task, all buttons, zippers, and other closing mechanisms must be completely closed. The top fastener near the worker's neck must be closed to minimize the chance that high temperature gasses could get behind the clothing and breach the protective characteristic. ASTM F1506 ensures that adequate layers of FR material protect fasteners on all FR-rated clothing from the extreme heat that could cause a worker to be burned.

The outer layer of the worker's protective clothing must be flame resistant. Any label or identifying patch must be made from the same protective material as the clothing. Any garment worn beneath the protective clothing should be made from flame-resistant fabric but must not be made from meltable or easily ignitable fabric. However, if the protective clothing will prevent an injury, it also will protect cotton underwear from igniting.

The head and face of a worker who must perform a task within the flash protection boundary are exposed to the potential arc flash. If the worker's head is within the flash protection boundary, he or she should select and wear protective equipment that will eliminate or minimize the risk of exposure to air or gasses that are very hot. Protective equipment that covers the head entirely is available. Face shields and balaclava are also available. Arc-rated goggles (**Figure 3-3**) might be acceptable for use with a balaclava (**Figure 3-4**). Equipment intended to protect a worker's head and face are assigned an arc rating that mirrors the arc rating assigned to FR clothing.

When a worker performs the necessary hazard/risk analysis, he or she must consider the location of the work

Figure 3-3 Goggles
Courtesy of Paulson Manufacturing

Figure 3-4 Balaclava (Head Sock)
Courtesy of PGI, Inc.

task in relation to his or her body. If the work task is near the floor, the worker must consider that hot air and gasses could get behind the protective equipment at the lower extremity. Injuries to a worker's legs are possible from the entrance of hot air and gasses under the leg protection on the clothing. The worker should ensure that protection for the bottom lower extremities prevents hot air and gasses from getting behind it. If the work task (and potential arcing fault) is at an elevated position, the risk of hot air and gasses getting behind the protective equipment near the worker's feet is reduced.

Protective Characteristics

Flame-resistant fabrics are tested as defined by ASTM F1506. Representative samples of the fabrics are subjected to the vertical flame test described in ASTM D6413, *Standard Test Method for Flame Resistance of Textiles*. This test determines that the fabric will not ignite and burn. The material is permitted to continue to flame for a period not exceeding two seconds after the test flame is removed and five seconds after an arc test. The char length in the vertical flame test is not permitted to exceed six inches. The test is conducted on new fabric and repeated on fabric that has been laundered 25 times.

Arc Rating

After the flammability of the fabric has been established, the fabric is then subjected to an arc test as described in ASTM F1959. An arc rating is established for each fabric sample when the F1959 test is conducted. The afterflame of the fabric must not exceed five seconds. Some fabrics exhibit an "afterglow" that exceeds five seconds. However, the afterglow is a property of the fabric and is not considered a flame. The product cannot exhibit any indication of melting or dripping when either the arc test or the vertical flame test is conducted.

The arc rating determines the protective characteristics of the fabric. When the product is sold to protect workers from arcing faults, clothing manufacturers are required to provide the arc rating on the label. Clothing manufacturers also are permitted to indicate the arc rating on the clothing surface, such as a shirt pocket or sleeve.

National consensus standards assign protective categories for arc ratings based on the protective nature of the equipment. **Table 3-4** illustrates the expected arc rating for

Table 3-4
Protective Clothing Characteristics

Hazard/Risk Category	Typical Protective Clothing Systems	Required Minimum Arc Rating of PPE [J/cm²(cal/cm²)]
	Clothing Description (Typical number of clothing layers is given in parentheses)	
0	Non-melting, flammable materials (i.e., untreated cotton, wool, rayon, or silk, or blends of these materials) with a fabric weight at least 4.5 oz/yd² (1)	N/A
1	FR shirt and FR pants or FR coverall (1)	16.74 (4)
2	Cotton underwear–conventional short sleeves and brief/shorts, plus FR shirt and FR pants (1 or 2)	33.47 (8)
3	Cotton underwear plus FR shirt and FR pants plus FR coverall, or cotton underwear plus two FR coveralls (2 or 3)	104.6 (25)
4	Cotton underwear plus FR shirt and FR pants plus multiplayer flash suit (3 or more)	167.36 (40)

Note: Arc rating is defined in NFPA 70E Article 100 and can be either ATPV or E_{BT}. ATPV is defined in ASTM F 1959-99 as the incident energy on a fabric or material to cause the onset of a second-degree burn based on the Stoll curve. E_{BT} is defined in ASTM F 1959-99 as the average of the five highest incident energy exposure values below the Stoll curve where the specimens do not exhibit breakopen. E_{BT} is reported when ATPV cannot be measured due to FR fabric breakopen.
Source: NFPA 70E-2004; *Standard for Electrical Safety in the Workplace*, Table 130.7(C)(11), Protective Clothing Characteristics.

various clothing categories. The protective nature of clothing categories applies to all arc-rated equipment. The clothing description in Table 3-4 is for illustration purposes only. Clothing manufacturers typically assign categories to various clothing combinations.

ASTM F1506 is a performance specification that identifies specific tests and acceptable results. The specification covers fabrics that are woven or knitted. The specification is not intended for use with other fabric construction.

In general, fabric is measured in weight per unit of area. For instance, ordinary denim jeans are on the order of 10 ounces per square yard. Fabric produced for use as FR clothing is also measured in ounces per square yard. As the weight per square yard of FR fabric increases, the thermal insulating ability also increases. In most cases, fabric from different clothing manufacturers is constructed from different textile products and blends of products from different textile manufacturers. The weight of the FR protective clothing is not an accurate indicator of the degree of protection provided by the clothing. The arc rating is the only reasonable indicator of the protection provided by the clothing.

Employers have a common practice of providing patches with an employee's name or a company logo. Employees are then encouraged to attach their name and company logo to their shirt or uniform. Any patch or logo attached to FR clothing also must be flame resistant. Patches that are flammable or meltable provide fuel that can ignite or melt and increase the effect of an arcing fault.

The protective characteristic of FR clothing depends on the surface of the fabric being clean and free from flammable material such as grease and oil. Wiping cloths should not be kept on a worker's body or clothing pocket when he or she is at risk of being exposed to a potential arcing fault. Just as a flammable patch adds fuel to the fire, grease and oil also add fuel to the fire. Wiping cloths that are kept in a pocket are likely to ignite and burn if exposed to an arcing fault. A burning wiping cloth in a pocket is likely to overcome the protection provided by the FR clothing.

Insulating Factor of Layers

Air is a relatively good thermal insulator. If two or more layers of FR fabric are used as part of a protective system, a layer of air is trapped between the layers. The air layer provides additional thermal insulation between a worker's skin and any potential arcing fault. Therefore, two layers of FR fabric provide greater protection than the arithmetic sum of the protective rating of the individual layers.

The basic idea of providing protection is adding thermal insulation between a worker and an arcing fault. As indicated above, a layer of air increases the thermal protection. FR clothing, then, should be loose fitting but not so loose as to interfere with the worker's movement. When the FR clothing is in contact with the worker's skin, thermal energy can be conducted through the fabric to the worker.

Fabric used to construct flame-resistant clothing provides only thermal insulation. It provides no protection from shock or electrocution. However, FR-protective equipment does not increase the chance of electrical shock unless the material is metalized or contains a conductive component, such as carbon. Although the reflectivity of a metalized garment would reduce the amount of energy conducted through the garment, the conductive garment would increase the chance of initiating an arcing fault. Some protective clothing contains a small amount of carbon to reduce the effect of static electricity. Small amounts of carbon embedded in the thread do not increase exposure to shock. Users should contact the clothing or fabric manufacturer for further details.

An FR-rated switching hood provides greater protection than a face shield. In addition to the thermal protection provided by an FR-rated hood, protection also is provided by a small quantity of air trapped in the hood. Should an arcing fault occur, the uncontaminated air inside the hood might be an advantage for a worker (see color insert, **Figure CP-3**).

A face shield provides protection from energy that is transmitted in the form of radiation. Energy that is transmitted by convection could flow behind the face shield and cause a burn. A face shield is less constricting than a hood and provides protection from flying parts and pieces that may be expelled by the blast component of the arcing fault (see **Figure 3-5**).

Figure 3-5   Arc-Rated Face Shield
Courtesy of Paulson Manufacturing

Supply and Provision

Employers may select different methods of providing FR clothing for employee use, depending on several variables. The electrical system in some facilities is simple and requires infrequent service. Other facilities have complex distribution and utilization systems that require several crews of workers to operate and maintain them. Some employers engage a contractor (another employer) to operate and maintain the electrical system. Other facilities might use explosive or flammable materials and require FR clothing for reasons other than exposure to electrical hazards.

Each employer must develop a strategy to ensure that employees are provided with arc-rated equipment when necessary. In some instances, the employer purchases multiple sets of FR clothing for each worker, and the worker is responsible for keeping the clothing clean and usable. The employer may choose to purchase multiple sets of FR clothing and store the clothing at various places across the facility where the clothing will be needed, thus ensuring that the workers have the protective clothing available when they need it. The employer develops a strategy to ensure the clothing is maintained. In other instances, the employer engages a uniform rental contractor to provide and maintain the FR clothing. Typically, the uniform rental contractor provides workers with multiple sets of FR clothing and exchanges dirty uniforms for clean uniforms at an established frequency. Such a service ensures that the FR clothing is maintained to protect the integrity of the FR properties.

Storage and Care

ASTM F1449, *Standard Guide for Care and Maintenance of Flame Resistant Clothing*, describes general recommendations for laundering and repair of FR clothing. Manufacturers are required to provide laundering instructions for FR clothing. FR clothing should be kept free from flammable contamination. Flammable contamination provides fuel for a fire that could ignite and burn on the surface of the clothing. FR clothing provides excellent thermal insulation for a short period of time. A fuel fire on the surface of FR clothing could last a relatively long time and result in an injury from the extended exposure to flame.

FR clothing is inherently resistant to ignition. Laundering has no impact on the thermal protective characteristics of FR clothing.

Some manufacturers provide clothing made from flammable material that has been treated with a chemical to resist ignition. These products commonly are called flame retardant. Treatment of the material with the chemical adds the flame-retardant property. Some of the desirable property is removed each time the product is laundered. Flame-retardant products must not be confused with flame-resistant products. Although the terms are similar, the characteristics are not similar.

▇ Selecting and Wearing Protective Apparel

Collecting and analyzing arc flash injury data are difficult tasks at best. It is clear that burns are the most common injury related to electrical energy. When OSHA promulgated Subpart S (29 CFR 1910.300-399) in 1990, the preamble indicated that burns were a major concern. The preamble suggested that burn injuries were the result of current flow through tissue, thermal burns that resulted from contact with a hot surface that resulted from current flow, and burn injuries that resulted from arcing contacts. Burns that result from an arcing fault were not a part of the puzzle, because no data existed. Burns that resulted from an arcing fault were typically treated as burns and recorded as burns, and the medical records contained no information relating the burn to electrical energy.

Internal records of major international corporations, however, suggest that arc flash burns are a significant percentage of the total number of electrical injuries. It has been estimated that the sum total of burn injuries (contact burns, current-flow burns, switching-arc burns, and arcing-fault burns) amounts to about 80 percent of all electrical injuries. Due to inadequate records, the accuracy of the estimate cannot be substantiated by publicly available injury data; however, burn injuries are the most numerous and are among the most serious injuries associated with electrical energy.

Preventing Burns from Clothing

Although FR clothing will prevent burns, anecdotal information suggests that the most severe injuries result from burning clothing or clothing that melted when exposed to an arcing fault. Faults are short circuits, which are sensed and cleared by overcurrent devices. The National Electrical Code® or the National Electrical Safety Code® specifies the maximum size and function of overcurrent devices. When the overcurrent devices are selected and installed as required by these codes, the duration of the arcing fault is controlled. When the duration of the fault is controlled, exposure is limited to clearing time of the overcurrent device. Any burn injury from the fault must occur within that short period of time.

When a worker's clothing ignites, however, the worker's skin is exposed to burning clothing for many seconds. The resulting injury is likely to be very severe. Some clothing is made from polyester, acetate, and similar materials that melt. When the clothing melts from exposure to an arcing fault, it contains a significant amount of thermal energy, which is transferred to the worker's skin and results in deep burns.

Wearing FR clothing, then, has three objectives:

1. To eliminate the chance of ignition of a worker's clothing
2. To eliminate the chance of a worker's clothing melting onto the worker's skin
3. To provide thermal insulation to reduce the risk of burns from direct exposure to an arcing fault

Workers should select and wear a protective apparel system that consists of protection for hands, arms, face, neck, trunk, legs, and feet that accomplishes these objectives.

When starting a campfire, kindling normally is used to get the fire started. Kindling is much easier to ignite than larger firewood. Although the ignition temperature of the wood does not change when chopped into kindling, the smaller, lighter pieces of kindling are much easier to ignite. Clothing made from cotton, wool, silk, or other natural fibers can ignite and burn.

Similar to attempting to start a fire with kindling, lightweight clothing is easier to ignite than heavyweight clothing. Blue jeans normally are made from textiles that weigh approximately 10 ounces per square yard, while shirts are made from textiles that weigh approximately 5 ounces per square yard. Therefore, blue jeans are less likely to ignite than shirts. The heavier weight of jeans material also provides greater thermal insulation. However, clothing that is not FR rated will ignite and burn.

The Protective System

The components of the protective system must be worn so that each mode of energy transfer is accounted for in the protection provided by each component. Energy transferred by convection suggests that protective components closest to the potential arc should extend over the adjacent component. For instance, the gauntlet of a worker's gloves should cover the end of the worker's sleeves covering his or her arms. Shirt collars should be buttoned to the top. If the worker is wearing a switchman's hood, the hood should extend well below the worker's neck. If the worker is wearing a face shield, the face shield should be long enough to provide protection for the worker's neck. Clothing should cover the worker's legs to below the top of the worker's shoes or boots. Shoes or boots should be heavy-duty, leather work shoes or boots. Cloth shoes such as sneakers must not be worn. If the potential arc source is near the worker's feet or lower extremity, the worker's body position should be such that exposure to the leg covering is not pointed toward the potential arc.

The outer layer of apparel worn by a worker should be flame resistant. If jackets, windbreakers, or sweatshirts are needed for warmth, these components should be covered by FR-protective equipment unless made from FR material. Normally, FR clothing has pockets just as ordinary street wear does. When a worker might be exposed to a potential arcing fault, he or she must remove everything from the pockets. Pencils, pens, bolts, nuts, screwdrivers, and other products should be removed from the FR protection. Flammable materials such as drying cloths, paper wipes, tissues, and writing pads should be removed from the pockets when the worker is exposed to a potential arc. Metal rings, earrings, hair adornments (including flammable hairnets), watches, and similar components should be removed when exposure to an arcing fault exists.

Workplaces such as food preparation areas, photography dark rooms, or pharmaceutical laboratories may require employees to work in "clean rooms," which may require employees to wear hair and head coverings as well as facial hair coverings. These products must be made from FR materials to avoid the risk of the coverings melting into hair or skin in the event of an arc flash. Clean rooms typically require employees to wear clothing that will not degrade the manufacturing product. These products normally are highly flammable (made out of disposable fabrics or paper) and must not be worn where the arc flash hazard exists. Some arc-rated products may be used instead of the flammable clothing, and they are available from some manufacturers.

▇ Selecting Flame-Resistant PPE

The flash hazard analysis must provide two critical pieces of information. First, the analysis must identify the flash protection boundary, that is, the point where unprotected skin would be injured from exposure to the thermal hazard. Second, the analysis must identify the amount of incident energy available at the distance to the potential arcing fault from the worker. Note that equipment should have a label that identifies incident energy available at a normal working distance.

Labels

The NEC requires that new installations of specific electrical equipment have a label installed in the field to warn workers when an arc flash hazard exists within the equipment. The requirement was first adopted with the issue of the 2002 edition of the National Electrical Code® and continues in the 2005 edition. The 2004 edition of NFPA 70E extracted the requirement for a warning label and included the language in Article 400.11.

The requirement does not specify a position location for the label; however, it must be clearly visible to any person who approaches the equipment. Neither the NEC nor NFPA 70E specifies the information that must be on the label. Therefore, the appearance and completeness of the information vary from one employer to another. However, the NEC suggests that the colors and font size for the warning label comply with ANSI standard Z535.4, *Product Safety Signs and Labels*.

Labels may be produced with a commercial software program (**Figure 3-6**) or printed from a standard color printer (**Figure 3-7**). The commercial labels define the arc flash boundary and the possible incident energy contained in the equipment. No additional information is required. However, both labels identify the shock boundaries in addition to the arc flash information. With the information on either label, a worker has all the information necessary to select equipment that will protect him or her from both arc flash and shock.

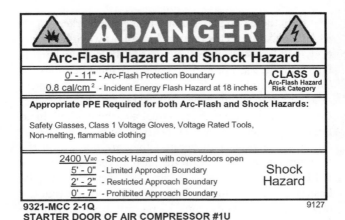

Figure 3-6 Commercial Label
This label is for illustrative purposes only and is not to be applied to any equipment.
Courtesy of GE Energy

> ⚠ **WARNING**
>
> ## Arc Flash and Shock Hazard
> ## Appropriate PPE Required
>
> _____ *Flash Protection Boundary*
> _____ Cal/cm² *flash hazard at* _____ **inches working distance**
> _____ **Shock hazard when cover removed**
> _____ **Inches** *Limited Approach Boundary*
> _____ **Inches** *Restricted Approach Boundary*
> _____ **Inches** *Prohibited Approach Boundary*
> _____ **PPE category**

Figure 3-7   Computer-Generated Label

With these two pieces of information, a worker can select protection from the thermal hazard. Any worker or any body part of a worker that is closer to the potential arcing fault than the flash protection boundary must be protected from the thermal hazard. For instance, if the flash protection boundary is four inches and the person's hand is closer than four inches from the potential arc, the worker's hand must be protected. However, other body parts are not required to be protected, because they are not within the flash protection boundary. If the flash protection boundary is 12 inches and the worker's hand and arm are within the flash protection boundary, both the hand and arm (but not other parts of the body) must be protected. Similarly, if the flash protection boundary is eight feet and the worker's entire body is completely within the flash protection boundary, the worker's entire body must be covered with PPE.

After the worker determines what part of his or her body must be covered by PPE, the worker must determine how much PPE is needed to avoid injury. Incident energy calculations estimate the amount of thermal energy that a worker's body would receive at a prescribed distance. In most instances, that prescribed distance is 18 inches; however, the required label should define the exposure distance included in the calculation. Workers should select PPE that has an arc rating at least as great as the incident energy indicated on the label. The arc rating of the PPE may exceed the incident energy on the label, but it must not be less than the amount indicated on the label.

PPE Configurations

PPE can consist of many different configurations, including various combinations of flame-resistant clothing and protective equipment (see **Figure 3-8** and color insert, **Figure CP-4**). A worker might choose to wear coveralls or a shirt and pants combination. The PPE might consist of cotton clothing and an FR lab coat, as long as the lab coat provides protection to the appropriate portion of the worker's body. The PPE might consist of a single layer, or it might consist of two or more layers. The PPE might consist of both natural fiber material and FR material. The manufacturer should be consulted to determine the overall rating of any multiple-layer protective system.

Workers must not wear clothing made from fabrics that melt, such as polyester, nylon, acetate, and similar products. Clothing containing blends of these products must not be worn unless the product is assigned an arc rating by the manufacturer.

Many undergarments are made of polyester, nylon, acetate, and similar products. If an undergarment melts, the melted fabric will destroy any skin tissue that it contacts. Workers who wear FR-protective equipment should

Figure 3-8a FR Shirt and Pants
Courtesy of Workrite Uniform Company

Figure 3-8b FR Jeans and Shirt
Courtesy of Tyndale Company, Inc

Figure 3-8c FR Long Coat
Courtesy of Workrite Uniform Company

Figure 3-8d Leather Gloves
Courtesy of Ironwear

Figure 3-8e   Complete Flash Suit
Courtesy of Salisbury Electrical Safety LLC

wear undergarments of cotton or flame-resistant fabric only.

An arcing fault converts electrical energy into other forms of energy. Flame-resistant products discussed in this section provide protection from the energy that is converted into thermal energy. All electrical arcs also convert some energy into pressure. The pressure wave created when an arcing fault is initiated produces a wave that results in a force. (That force, known as blast, is discussed in Chapter 4.)

Flame-resistant PPE offers little protection from forces associated with the pressure wave. As the electrical energy in the arc increases, both the thermal and blast forces

Figure 3-9   Arc-Rated Bra
Courtesy of ArcWear

increase. Arcing faults that produce incident energy greater than 40 calories per square inch produce pressure waves that may cause injury to a worker, regardless of the amount of FR protective equipment the worker is wearing. If the flash hazard analysis indicates that the incident energy is 40 calories per square inch or more, the work task should not be performed while the circuit or equipment is energized.

Clothing Designed Especially for Women

Until a few years ago, FR clothing was made in sizes for male workers only, although women might wear it. Several manufacturers such as ArcWear, Bulwark, Tyndale, and Workrite now offer FR and arc-rated clothing, including both outerwear and underwear, in standard sizes for women (see **Figure 3-9** and color insert, **Figure CP-5**).

Some companies also make arc-rated underwear for men that contains no metallic components. Men can wear cotton underwear under their arc-rated clothing with no danger of it melting. However, some women's underwear is made from polyester or similar materials that could melt in an arc-flash event, and some bras contain metal under-wires, hooks, or clips that should not be worn (even under FR clothing) if there is potential for exposure to the arc-flash hazard.

References

ANSI Z535.4, *Product Safety Signs and Labels*. New York, NY: American National Standards Association, 1998.

ASTM D6413, *Standard Test Method for Flame Resistance of Textiles*. Conshohocken, PA: American Society of Testing and Materials, 1999.

ASTM F1449, *Standard Guide for Care and Maintenance of Flame-Resistant Clothing*. Conshohocken, PA: American Society of Testing and Materials, 2001.

ASTM F1506, *Standard Performance Specification for Flame-Resistant Textile Materials for Wearing Apparel for Use by Electrical Workers Exposed to Momentary Electric Arc and Related Thermal Hazards*. Conshohocken, PA: American Society of Testing and Materials, 2002.

ASTM F1959, *Standard Test Method for Determining the Arc Rating of Materials for Clothing*. Conshohocken, PA: American Society of Testing and Materials, 2005.

Doughty, Richard L., et al. "Predicting Incident Energy to Better Manage the Electric Arc Hazard on 600V Power Distribution Systems." PCIC Paper PCIC-98-36. Paper presented at the Forty-Fifth Annual Conference of the IAS/IEEE Petroleum and Chemical Industry Committee, Indianapolis, Indiana, September 28–30, 1998.

Doughty, Richard L., et al., "Testing Update on Protective Clothing and Equipment for Electric Arc Exposure." PCIC Paper PCIC-97-35. Paper presented at the Forty-Fourth Annual Conference of the IAS/IEEE Petroleum and Chemical Industry Committee, Banff, Alberta, September 15–17, 1997.

IEEE Standard 1584, *IEEE Guide for Performing Arc-Flash Hazard Calculations*. New York, NY: Institute of Electrical and Electronics Engineers, 2002.

The National Electrical Code (ANSI/NFPA 70). Quincy, MA: National Fire Protection Association, 2005.

The National Electrical Safety Code (ANSI/IEEE C2). New York, NY: Institute of Electrical and Electronics Engineers, 2007.

NFPA 70E, *Standard for Electrical Safety Requirements for Employee Workplaces*. Quincy, MA: National Fire Protection Association, 2004.

U.S. Department of Labor. Occupational Safety and Health Administration. OSHA Regulations 29 CFR 1910.300-399, Subpart S, "electrical." Washington, DC.

U.S. Department of Labor. Occupational Safety and Health Administration. OSHA Regulations 29 CFR 1910.269, Subpart R, "electric power generation, transmission, and distribution." Washington, DC.

Protection from Arc Blast

As discussed in the previous chapter, electrical current flowing through a conductor generates heat. In the case of an arcing fault, current flow is through contaminated air. The flowing current produces visible plasma in the space between the expected electrical conductors. It is inside the plasma that the temperature reaches many thousands of degrees. Overcurrent devices that are selected and installed as defined by consensus standards normally operate in a very short period of time, which limits the duration of the plasma.

Copper is a common electrical conductor material; it melts when heated to about 2000°F and boils when heated to 4703°F. As noted previously, the temperature of the plasma reaches an extremely high temperature. An arc temperature of 16,000° to 18,000°F is quite common. Of course, one terminal point of the electrical arc (and plasma) is the electrical conductor. If the conductor is copper and the temperature of the plasma is 16,000°F, the copper will melt, boil, and turn to copper vapor.

■ Arc in Open Air or Enclosure

Air that ordinarily surrounds conductors is heated by being in the proximity of the electrical arc. As the temperature of the air increases, the air expands and requires significantly more space to keep the pressure constant. If the arcing fault is contained in an enclosure, the air is not allowed to expand. As the temperature rises, the air continues to try to expand, which increases the pressure in the enclosure.

When copper changes from a solid to a liquid, it expands slightly. When the liquid copper boils, it changes from liquid to vapor, and the material expands many thousands of times. One square inch of copper, as a solid, requires several thousand square inches as a liquid at standard pressure. If the copper vapor is constricted so that it cannot expand rapidly enough, the result of the vaporizing liquid copper will be an increase in pressure. If the arcing fault is within a full or partial enclosure, the increase in pressure will be significant.

It would seem that an arc in open air does not restrict the expanding heated air and copper vapor. However, the transfer of thermal energy from the arc plasma to the copper and surrounding air is much faster than the ability of nearby air to move away and allow the expanding materials to fill the space. The interface between air at atmospheric temperature and pressure and the expanding gasses is a point where the pressure increases very rapidly, creating a wave front. A lightning stroke is an example of an electrical arc in open air. The sound associated with the lightning bolt is the pressure wave created when the current flowing in the discharge heats the air. The wave front moves very fast and approaches the speed of sound.

If the arcing fault is contained in an enclosure, the heated air and copper vapor are prevented from expanding by the enclosure. If the fault continues to arc, the pressure within the enclosure also continues to increase. When the pressure inside the enclosure exceeds the ability of the enclosure to contain the pressure, the enclosure ruptures and relieves the pressure. Pressure is mechanical energy that has been converted from thermal energy, which in turn has been converted from the electrical energy.

The increasing pressure can force parts and pieces off the equipment and fling them away from the arc or equipment. The expanding gasses contain metal vapor that cools and condenses as it flies through the air or condenses when the metal vapor contacts an object that is much cooler, such as a person's skin or clothes.

The amount of force associated with an object that is propelled away from an arcing fault depends on the momentum of the moving object. Momentum depends on the weight of the object and its velocity (speed). As the speed of an object increases, the potential force associated with the object also increases. The speed of the object depends on the amount of pressure generated during the arcing fault. Anecdotal information suggests that a fault inside a motor terminal box can rupture the iron enclosure and fling a portion of the terminal box across a room and even penetrate another enclosure at a distance of 20 feet. In other instances, the pressure generated in an arcing fault has destroyed concrete block walls.

▓ Pressure Wave

The force applied to an object by the moving wave front depends on the difference in pressure at the leading edge of the wave front and the other side of the object, in addition to the physical size of the obstruction. If the obstruction is a person standing in front of the arcing fault, the size of the obstruction is about three square feet. The force exerted on the person, then, is the pressure in pounds per square foot multiplied by the number of square feet. If the pressure at the person's back is atmospheric pressure, the person will feel a force that is equal to the difference in pressure in pounds per square inch from the front of his or her body and the back of the same person (see **Figure 4-1**).

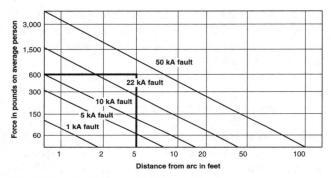

Figure 4-1   Arc Blast Pressure on a Typical Human Body
Adapted from Ralph H. Lee, "Pressures Developed by Arcs." *IEEE Transactions on Industry Applications*, vol. 1A-23, no. 4, 1987. Piscataway, NJ: IEEE.

If the object obstructing the expanding gasses is an equipment door, the pressure in pounds per square inch multiplied by the area of the door in square inches determines the amount of force exerted on the door. When the force from the expanding gasses exceeds the retaining force of the latches, or when the internal force bends and deforms the enclosure to release the latches, the door will open and expose a worker standing in front of the door to the wave front. Although the pressure in the wave front decreases with time and distance, flying parts and pieces maintain significant momentum. The only slowing force on a part that is ejected by an electrical arc is the resistance offered by the air until the object contacts a physical obstruction, such as a person's body.

▊ Protective Materials

Leather products are intended to avoid abrasion and penetration. Although leather provides sound protection from abrasion and from penetration by parts that are static, moving parts and pieces ejected by an arc blast will be impeded only partially. Some FR clothing contains material (such as Kevlar®) that normally is used to stop small flying objects such as bullets. If the FR clothing successfully stops the moving part, the energy contained in the moving object will be transferred to the person. Face shields, safety glasses (spectacles), and viewing windows in hoods are intended to provide protection from impact. However, these products are not tested in an environment where objects of significant size and momentum might exist. Rubber products worn for protection from electrical shock provide a layer that can resist penetration and abrasion. Again, however, energy contained in the flying parts and pieces is transferred to the person, even if the rubber protection is not penetrated.

The bottom line is that no products exist that can protect a person from the blast effects of an arcing fault. Products worn for protection from other electrical hazards can provide some resistance to penetration, but they do not absorb the kinetic energy in the momentum of flying parts and pieces. Only two options exist. First, the equipment must be placed in an electrically safe work environment. Second, there must be sufficient distance from the source of the potential arc.

Reference

Lee, Ralph H. "Pressures Developed by Arcs." *IEEE Transactions on Industry Applications*, Vol. 1A-23, No. 4, July/August 1987. Piscataway, NJ: IEEE, 1987.

General Protection from Electrical Injuries

Most people think of PPE only as clothing or rubber insulating products. This text has already discussed PPE that guards against the hazards associated with shock and arc flash, even though some of these items might not normally be thought of as personal protective equipment. This chapter considers general PPE that can help prevent injuries against those and other hazards in electrical work:

- Voltmeters
- Hard hats
- Spectacles (safety glasses)
- Face shields and viewing windows
- Voltage-rated hand tools
- Safety grounds (clusters)

Voltmeters

For a worker to be safe from an electrical incident, the most important thing that he or she must know is whether an electrical conductor is energized. If the electrical conductor is energized, the next most important bit of information is the level of the voltage that is present. Only a voltmeter can determine whether a conductor is energized. Although the voltage-detecting device is not worn or installed, as is the case with other PPE, the construction and integrity of the device are critical. Should a voltmeter fail while in direct contact with an exposed energized conductor, the result

117

could be an arcing fault. Should a voltmeter fail to accurately sense the presence of voltage, the result could be electrocution. A worker can sense the presence of voltage only by using an effective voltage detector.

Many different types of voltmeters and voltage detectors are available for purchase (see color insert, **Figures CP-6** and **CP-7**). In most instances, the devices work well within the intended parameters. To avoid misapplication of the devices, workers must be familiar with the intended purpose of the device.

The Home Environment

Voltmeters are available from many manufacturers and in many different sizes and configurations. Some devices are intended for use by a hobbyist working on circuits with limited capacity. Usually, these meters are sold in home improvement and hobby stores. Provided the capacity of the source of electrical energy is limited to a level that will not provide sufficient energy to generate an electrical hazard, devices from hobby shops may be used safely.

Home improvement stores stock voltmeters that are likely to be used on circuits in the home environment. In a home environment, the system capacity could be sufficient to sustain an arcing fault. Electrical circuits in the range of 120 volts to 240 volts are high enough to result in an electrocution. Although the capacity of a residential system is much smaller than a commercial or industrial system, the integrity of a voltage-measuring device is equally important.

Selection and Use

Users should select a voltmeter that has a range commensurate with the expected circuit voltage. If the voltmeter is intended to verify the absence of voltage, users should select a device intended for direct contact. If the level of voltage is not known, users should select a meter that has auto-ranging capability, but users must ensure that the circuit voltage is lower than the maximum range of the meter.

When a voltage-detecting device is being used, electrical injuries occur for one or more of the following reasons:

- The voltmeter was misapplied or misused.
- The voltmeter was selected improperly.
- The indication was misunderstood.
- The leads came out of the meter and touched the grounded enclosure, resulting in a short circuit.
- The user's hand slipped off the end of the probe and contacted the energized conductor.
- An internal failure occurred, and the meter exploded (see color insert, **Figure CP-8**).

When the probes from a voltmeter are in contact with energized conductors, the circuit under test experiences an additional circuit element. Current flows between one voltmeter probe and the other. The amount of current depends on the internal impedance of the voltmeter. Current is measured on a scale (either analog or digital) that is graduated in volts. In solenoid-type devices, the amount of current flow exerts a known magnetic force on the solenoid. The solenoid movement is graduated in voltage. For the voltmeter to function effectively, the integrity of the current path through the measuring element is critical. The internal current path includes the probes, the plug in the case that accepts the probes, and internal components. In some cases, current flows through the switch used for changing scale. The switch easily could be set to the incorrect position, which would destroy the meter and could cause an arcing fault.

Devices with low internal impedance tend to discharge induced or static voltage. High internal impedance devices may measure all voltage, including induced and static voltage. Workers must understand the relationship of internal impedance to measuring voltage and the internal impedance of the current path.

Some voltage-detecting devices do not require direct contact with an exposed energized conductor (see color insert, **Figure CP-9**). Non-contact voltage, detecting devices sense the presence of an electrostatic or electromagnetic field. These devices look for inductive or capacitive coupling between the device and the conductor in question. Normally, non-contact voltage detectors provide an audible signal that a voltage is present. Some devices also provide a visual signal. Depending on how the non-contact voltage detector works and the physical construction of the conductor in question, a null point may exist along the conductor's linear direction. These devices provide an initial indication of the presence of voltage, but they should not be used to determine the absence of voltage.

Duty Cycle

Depending on the construction of the device, the manufacturer might assign a duty cycle to the device (see color insert, **Figure CP-10**). A duty cycle is intended to prevent the device from overheating. Solenoid-type voltage-detecting devices are constructed by wrapping a small wire around a core to form the coil of a solenoid. The small wire usually has varnish for insulation. To avoid overheating the insulating varnish, a duty cycle is assigned to permit the coil to cool. A rule of thumb is that a solenoid voltage detector should be used for no more than 15 seconds without permitting the device to cool for at least another 15 seconds. Manufacturers define the duty cycle on the label attached to the equipment.

Although a voltmeter may be used to troubleshoot a circuit, the voltmeter is a safety device. A voltmeter provides crucial information for a worker to evaluate his or her exposure to an electrical hazard. The voltmeter is PPE in the same vein as safety glasses (spectacles).

Electricians sometimes keep a voltmeter in their toolbox with screwdrivers, socket wrenches, pipe wrenches, and so on. Workers then tend to view their voltmeter as a hand tool that is used to troubleshoot a circuit, giving little thought to the fact that their lives depend on the integrity of their voltmeters. Some workers carry voltmeters that were purchased at a hobby store because those devices may be cheaper. Still other workers carry a shirt pocket version of a non-contact voltage detector. These approaches may prevent workers from accurately sensing the presence or absence of voltage in a circuit, which is the first step in preventing injury.

UL 1244 Requirements

The worker should purchase a voltmeter that complies with specifications relevant for the particular intended use of the voltmeter. The national consensus standard covering voltmeters is UL 1244, *Standard for Electrical and Electronic Measuring and Testing Equipment*. Requirements defined in this standard address issues that result in injuries associated with the construction of voltmeters that were known when the standard was promulgated. For instance, UL 1244 requires that the banana plugs that connect the leads to the voltmeter be shielded to prevent contact with a grounded surface in case either of the plugs slips out of the receptacle. The standard requires that probes contain a knurled section near the end to help prevent a worker's hand from slipping and contacting the energized conductor. The standard requires that the meter be designed such that the mode selector switch cannot be a part of the active circuit. The standard requires adequately rated fuses, eliminating the risk of initiating an arcing fault from component failure.

Only voltage-detecting devices that are evaluated and comply with UL 1244 may display the UL label. Unless products are so marked, they do not comply with the standard's requirements.

Electrical systems are becoming increasingly complex. The amount of equipment that generates transient voltage spikes on a system is increasing. A transient voltage spike results from a static discharge in a lightning strike. Transients also might be the result of switching inductive loads. Normally, voltage spikes are a few microseconds in duration but can involve many hundreds of amperes. In some areas of North America, lightning discharges are common, especially in the spring and summer months. When a transient spike is in an electrical circuit, any equipment electrically connected to that circuit must be capable of handling the spike. Transients can destroy a voltmeter that happens to be in contact with the electrical conductor simultaneously. Transients contain less energy as the distance from the source of the transient increases.

Static Discharge Categories

ANSI/ISA Standard S82.02.01, *Electrical and Electronic Test, Measuring, Controlling, and Related Equipment, General Requirements*, and similar international standards establish a rating system for voltmeters. Voltmeters and other measuring instruments are assigned to categories. The categories differentiate the ability of the devices to handle the energy in a transient condition (such as a spike in voltage caused by lightning or switching). The categories are established by the location in the circuit between the source of electricity and the point where the device will be used.

Voltmeters that can be used in the point of generation and transmission (where most energy is available) are Category (CAT) IV devices. As the distance from the generator or transmission line to the point of use increases, the assigned category decreases. CAT III devices can be used safely on distribution level circuits such as motor control centers, load centers, and distribution panels. CAT II devices can be used safely on receptacles and utilization circuits. CAT I devices are intended for use on electronic equipment and circuits. As the category decreases (from IV to I), the ability of the device to resist damage from transient over-voltage decreases. Therefore, CAT I devices are less likely to survive a transient over-voltage (spike). For an illustration of category uses of voltmeters, see **Table 5-1** and **Figure 5-1**.

In the international community, voltmeters are assigned to categories according to their ability to function in various environments where transient currents are expected. The international standard that covers transient categories for voltmeters is IEC 61010, *Safety Requirements for Electrical Equipment for Measurement, Control, and Laboratory Use*. The transient categories assigned by IEC 61010 are CAT I, CAT II, CAT III, and CAT IV. These transient category ratings align with ratings established by the ANSI/ISA system. Voltmeter ratings based on ANSI/ISA S82.02.01 are equivalent to the ratings based on IEC 61010.

Table 5-1
Transient Ratings of Voltmeters

Category Rating	Suggested Use
I	Electronic equipment and circuits
II	Receptacles and utilization circuits
III	Distribution level circuits such as motor control centers, load centers, and distribution panels
IV	Generation and transmission circuits
Source: ANSI/ISA Standard S82.02.01	

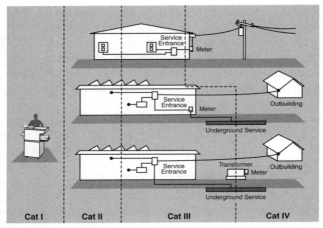

Figure 5-1   Illustration of Transient Category-Rated Meters

Purchase

Voltmeters provide information that is critical to preventing injuries. Electricians and technicians often keep their voltmeters in the same toolbox where they keep socket sets, screwdrivers, pipe wrenches, and other hand tools. As the toolbox is bumped and dropped, the critical voltmeter is subjected to the same physical forces as the metal tools. The voltmeter case may be damaged or internal components may be damaged by the physical shock. Voltmeters that will be subjected to physical shocks should meet a specification that provides for protection from the physical abuse. The authors recommend that the purchase order for voltmeters require the device to have a UL label indicating compliance with UL 1244. The authors also recommend voltmeters for use in an industrial setting be assigned a CAT IV rating, as determined by ANSI S82.02.01 or IEC 61010.

Inspection

Users should inspect each voltmeter each day before it is used. The user must always ensure that the voltmeter is functioning normally before conducting the test and then verify that the voltmeter is still functioning normally after conducting the test. The inspection should verify the following information:

- The protective fuse is good.
- The case/enclosure is free from cracks and not otherwise broken.
- The readout is clear and legible.
- The insulation on the leads is complete and undamaged.
- The shroud on the banana plug is complete and undamaged.
- The finger guards are in place.
- The retractable probe covers are in place and functional.
- Each lead is continuous.

Hard Hats

Hard hats are intended to provide head protection from falling objects and bumping and to prevent the head from contacting energized conductors (see color insert, **Figure CP-11**). Gravity usually causes falling objects to travel straight down from an elevated position. The forces of falling objects are applied directly to the top of the protective helmets and distributed onto the worker's head by the helmet's suspension system. Sometimes a falling object strikes a fixed object or structure and is directed onto the worker's head from the side. The helmet assembly must distribute the kinetic energy contained in the falling object in this instance as well.

The test setup for the different types of exposure to hazards of falling objects or bumping must account for the various directions of the forces applied to the hard hat. National consensus standards use types to differentiate these exposures. Type I helmets are tested to distribute downward forces adequately. Type II helmets are tested to distribute both downward and lateral forces adequately. Suspension systems may have four or six points of support for the hard hat (see **Figure 5-2**). Six-point systems provide greater distribution of the impact.

ANSI Z89.1 Requirements

One main ANSI national consensus standard now offers guidelines for head protection: ANSI Z89.1, *Requirements for Industrial Head Protection*. ANSI does not write or publish standards. Instead, it selects and designates standards-developing organizations (SDOs) as secretariat organizations for standards. ANSI assigned the 2003 edition of the standards covering protective helmets to the International Safety Equipment Association (ISEA).

Figure 5-2   Hard Hat Suspension System
Courtesy of Salisbury Electrical Safety LLC

Previously, the American Society of Safety Engineers (ASSE) had served as the secretariat for standards covering protective helmets. As a result of the change in secretariat, some structural changes were made to the affected documents. One change eliminated ANSI Z89.2, *Safety Requirements for Industrial Protective Helmets for Electrical Workers,* and integrated the requirements for protection from electrical shock into ANSI Z89.1, which is the standard that covers impact and penetration. Other changes modified the class designated for electrical shock protection from Classes A, B, and C to Classes E, G, and C (see **Table 5-2**).

The change in secretariat for the national consensus standards occurred several years after applicable OSHA standards were issued. OSHA references to ANSI standards refer to editions of the national consensus standards that were effective when the regulation was last promulgated. ANSI Z89.2 was affected by the infrequent revision of OSHA standards. Although ANSI Z89.2 has been eliminated, it is still referenced in 29 CFR 1910.6. Therefore, users should ensure that they locate and use protective equipment that complies with the latest edition of the applicable national consensus standard.

Where the hazard analysis indicates a chance for an elevated object to fall, employers are required to ensure that workers wear head-protective equipment. Basic require-

Table 5-2
Classes of Hard Hats

Class	Maximum Voltage Rating	Equivalent Older Designation
E	20 kV	B
G	2.2 kV	A
C	No rating	C

Note: Class C hard hats are conductive.
Source: ANSI Z89

ments for protection from falling objects are defined in ANSI Z89.1. If the hazard analysis indicates a chance that the worker's head might be exposed to electrical shock as well as falling objects, current OSHA regulations indicate that helmets also must comply with ANSI Z89.2-1971. ANSI Z89.2 is identified in 29 CFR 1910.6 as incorporated into the OSHA standards by reference.

OSHA Requirements

In 29 CFR 1910.132, OSHA requires employers to ensure that workers wear hard hats that provide protection from falling objects. Employers must execute and document a hazard analysis to decide what type of head protection workers should use. If the potential injury is limited to a worker bumping his or her head on an obstruction, the head protection need only consist of bump caps, which are lightweight and less restricting than heavier hard hats. However, use of bump caps is restricted to applications where falling objects are not anticipated.

A requirement for protective helmets that reduce the electrical shock hazard is cited in 29 CFR 1910.135(a)(2). In 1910.268(j)(1); Class B head protection is required when a worker might be exposed to high-voltage electrical contact. [Note that "Class B" was a designation in the now superseded ANSI Z89.2. The current ANSI Z89.1 calls this head protection Class E (see Table 5-2).] In 1910.335 (a)(1)(4), OSHA requires that workers wear head protection where danger from shock or burns is possible. The employer is responsible for ensuring that employees comply with these requirements.

ANSI Z89.1- 2003 assigns helmets to a class as defined by voltage. Table 5-2 indicates the test voltage for assigned voltage classes. Note that Class C hard hats are conductive and should not be worn by workers who are or may be exposed to an energized uninsulated electrical conductor.

Selection and Use

Hard hats are available in many different colors. Some employers provide workers in each discipline or craft with a unique color. In other instances, the color of the worker's hard hat indicates contract workers or specific contractors. In either instance, however, the hard hat should provide the impact protection and electrical shock protection defined in ANSI Z89.1-2003. In the opinion of the authors, the best alternative is for all workers to wear head protection that is rated as Class E protective hard hats.

Hard hats are made from various moldable materials. Polyethylene is common, because it can be molded easily into the required form and offers excellent resistance to abrasion and breaking. Hard hats may be constructed from other materials, provided the completed assembly meets the impact, penetration, and insulating characteristics defined in ASNI Z89.1.

Current national consensus standards for protective helmets define tests for impact, penetration, and electrical conductivity. However, ignition and flammability characteristics for protection from arc flash events are not addressed. Although ignition and flammability are important for use by firefighters, exposure to an arc flash event is not currently considered important for construction of hard hats. A hard hat made from polyethylene and similar materials could ignite when exposed to an arcing fault and should be covered by a flame-resistant hood or similar product if the hazard/risk analysis indicates that the worker may be exposed to an arc flash event.

Workers sometimes put decals on their outer surface of their hard hat. If the hard hat is Class E or G (i.e., rated for use near exposed energized electrical conductors), the decal should be nonconductive. Metal or conductive decals could be the cause of a short circuit and initiate an arcing fault.

Chinstraps and other attachments are available for hard hats. Adding an attachment might provide additional utility; however, the attachment might increase the potential for damage from a different hazard. For instance, a chinstrap will prevent the helmet from falling off the head, but can increase the amount of fuel if it should be ignited in an arc flash event.

Purchase specifications for hard hats must identify the characteristics of the desired helmet. It must indicate size (small, standard, or large), type (Type I or Type II), and color. When a hard hat is purchased, the suspension system must be purchased also. The suspension system must match the type and size of the hard hat. Incorrect application of the suspension system negates the protective characteristics of the protective helmet. Suspension systems and protective helmets from different manufacturers must not be intermingled. Hard hats are sold through distributors and local safety equipment suppliers.

Spectacles (Safety Glasses)

National consensus standards define performance criteria, testing requirements, and required marking for safety glasses. ANSI Z87.1, *American National Standard for Occupational and Educational Eye and Face Protection Devices*, indicates that safety glasses can be either basic impact or high impact. Each level of impact protection requires different test setup and acceptance criteria. Purchase orders must specify if basic or high-impact safety glasses are required. The terms "eye glasses" and "safety glasses" are commonly used to mean an assembly of lenses together with a supporting frame that is worn to correct a person's vision or protect a person's eyes from impact (see **Figure 5-3**). The

Figure 5-3 Safety Glasses
Courtesy of Ironwear

term "spectacles" is used in consensus and regulatory standards to have the same meaning. Spectacles, however, excludes coverall safety glasses and goggles.

In 29 CFR 1910.133(b)(2), OSHA suggests that protective equipment for eyes must comply with the requirements of ANSI Z87.1. This standard defines performance requirements that guide the construction of frames and lenses. By setting performance requirements, frames and support structures may be made from several different materials and with different construction, provided the overall assembly complies with the specified performance criteria.

The purpose of this book is to address electrical PPE. Safety glasses, as such, are not electrical PPE. However, safety glasses could fall from a worker's face and initiate an arcing fault if the protective equipment is conductive. In some instances, the supporting frame for safety lenses is made of metal and is conductive. Unrestrained spectacles (safety glasses or otherwise) must not be worn when a worker is performing work on or near exposed live parts.

■ Face Shields and Viewing Windows

Every work task should begin with a hazard/risk analysis. Workers cannot select and wear adequate PPE until and unless all potential hazards have been identified. Unless workers are familiar with the nature and limits of protective characteristics for available PPE, the workers cannot choose the protective equipment with confidence. Workers, supervisors, and managers must be trained to recognize the protective limits provided by equipment that meets national standards.

ANSI Z87.1 Requirements

Face shields might provide protection from several hazards. In some instances, a face shield is necessary to provide protection from impact. Grinding and cutting tasks usually generate flying objects that could injure a worker's face or eyes. A face shield that meets the impact and penetration requirements defined in ANSI Z87.1, *American National Standard for Occupational and Educational Eye and Face Protection Devices*, will provide adequate protection in this situation.

When workers are required to work with and handle liquid chemicals, a face shield will provide protection from occasional splashes that might occur. If the face shield is impervious to attack from the chemical, the face shield will provide adequate protection. If the face shield used for protection from a chemical spill also meets the impact requirements of ANSI Z87.1, the same face shield may be worn for protection from both hazards.

The same face shield that provides chemical and impact protection will filter some wavelengths of energy in the electromagnetic spectrum. Face shields that meet the requirements of ANSI Z87.1 provide protection from nonionizing radiated energy. Some ultraviolet energy is filtered as well. However, face shields and other equipment covered by ANSI Z87.1 do not provide protection from infrared energy or thermal energy (see color insert, **Figure CP-12**).

An arcing fault generates significant thermal energy in the form of heated air and gasses that rush toward the worker's face at speeds approaching the speed of sound. Molten copper and parts and pieces of the equipment might be rushing toward the worker's face at near the speed of sound. Significant electromagnetic energy is generated in an arcing fault and is radiated toward the worker's face at the speed of light. Workers should wear protective equipment that provides protection from all of these destructive components. An ordinary face shield intended for protection when using a grinder will not protect a worker from exposure to hazards associated with an arcing fault.

Direct contact between an exposed energized conductor or circuit part and a worker's face is unlikely. However, a worker's face is very likely to be exposed to the heated gasses and molten metal associated with an arcing fault. The worker's face and head must be protected from the destructive nature of these hazards. Therefore, the face shield must both meet impact requirements and provide adequate protection from the thermal hazard.

ASTM F2178-02 Requirements

ASTM F2178-02, *Standard Test Method for Determining the Arc Rating of Face Protective Products*, defines a testing method to establish a rating for face shields and viewing windows where the product will be exposed to an arcing fault. The rating system established in this standard suggests that products should be assigned an arc rating that essentially mirrors ratings for clothing defined in ASTM F1506 *Standard Performance Specification for Flame-Resistant Textile Materials for Wearing Apparel for Use by Electrical Workers Exposed to Momentary Electric Arc and Related Ther-*

mal Hazard. The arc rating should be based on the arc thermal performance value (ATPV) or the threshold breakopen energy (E_{BT}). Manufacturers assign arc ratings based on internal tests. No third-party testing process exists for establishing an arc rating.

Face shield selection should be based on the expected available incident energy that is identified when the hazard/risk analysis is performed. A face shield should be selected that has a rating greater than the expected incident energy exposure. When face protection is an integral part of another item of protective apparel, the viewing window should have the same rating as the other protective apparel. Consult the manufacturer of the protective apparel.

A face shield provides an edge on each side of the head of the person wearing it. A face shield obstructs the movement of the air and gasses rushing from the arcing fault during an exposure. Objects such as molten metal will be stopped by the obstructing nature of the face shield. However, the obstructing nature and curved shape of a face shield also obstruct the movement of air and gasses rushing from an arcing fault. If the worker's head is positioned such that the rushing air is not directed at the front of the face shield, the curvature of the face shield could result in a vacuum being generated behind the face shield by the air and gasses rushing past. Lift is generated under an airplane wing by the same action (see **Figure 5-4**).

All parts of a worker's body that are within the flash protection boundary must be protected from the effects of the potential arcing fault. If the worker's entire head is within the flash protection boundary, a face shield will not provide adequate protection. Only a switching hood with an adequately rated viewing window will provide the necessary protection in the case where worker's head is within the flash protection boundary.

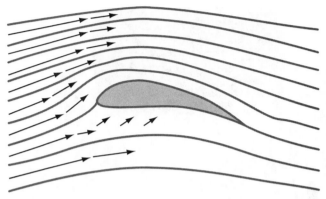

Figure 5-4   Illustration of Lift

Voltage-Rated Hand Tools

In 29 CFR 1910.333(c)(2), OSHA suggests that voltage-rated tools must be used for all work where the hand tools might make contact with an exposed energized conductor. NFPA 70E also contains a similar requirement. ASTM F1505, *Standard Specification for Insulated and Insulating Hand Tools*, defines construction and testing of hand tools that are rated for use on circuits that are 1000 Vac (Volts of Alternating Current) or 1500 Vdc (Volts of Direct Current).

The term "voltage-rated hand tools" refers to both insulated and insulating hand tools. Insulated hand tools are constructed from conductive material or components and have electrical insulation applied on the exterior surface. Insulating hand tools are constructed from nonconductive material. Insulating hand tools might have metal or other conductive inserts for reinforcement but, essentially, the

Figure 5-5   Voltage-Rated Hand Tools
Courtesy of Salisbury Electrical Safety LLC

tool must be constructed of nonconductive material (see
Figure 5-5).

Voltage-rated hand tools are not intended to serve as
primary protection from shock or electrocution. Although
insulated hand tools might provide adequate shock protection for circuits below 1000 Vac, workers must select and
wear PPE that provides protection from shock and arc flash
without considering the hand-tool rating. Although insulated and insulating hand tools include insulation from
electrical sources of up to 1000 Vac, the primary function
of the insulation is to reduce the risk of initiating an arcing
fault.

The insulating coating of insulated hand tools may
consist of a single layer, but normally it consists of two layers of contrasting colors. The interior layer provides 100
percent protection from shock to the full rating of the tool.

The contrasting color of the exterior layer provides a method for inspecting the tool for damage. Any cut or abrasion that exposes any of the inner layers constitutes significant damage and suggests that the tool should not be used. The damaged tool should be replaced with a new tool.

Workers should visually inspect each voltage-rated tool before each use. If the interior layer is visible, the tool should be discarded. The manufacturer must mark voltage-rated hand tools as follows:

- Manufacturer's name or trademark
- Type or product reference
- Double triangle symbol (see **Figure 5-6**)
- 1000-V
- Year of manufacture

ASTM F1505 Requirements

ASTM F1505 defines the testing of voltage-rated hand tools. Unless specifically approved for use at low temperatures, the tool may be used in ordinary atmospheric temperature ranging from 20° to 70°C. The standard requires the mechanical integrity of the tools to meet the same requirements as nonrated hand tools.

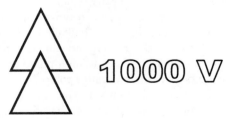

Figure 5-6   Double Triangle Symbol

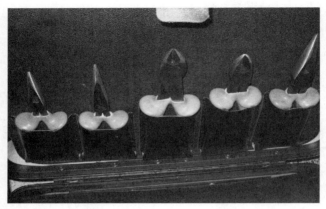

Figure 5-7   Sets of Insulated Hand Tools

Only hand tools that meet the requirements of ASTM F1505 should be purchased. A storage toolbox should be purchased with the tools (see **Figure 5-7**). The tools should be kept in the storage toolbox when not in use. The tool-box, with the tools, should be stored in a location that is clean and dry.

Safety Grounds (Clusters)

A deenergized electrical distribution circuit could be reenergized by several different means, which creates an unsafe condition. If the conductors are on cross arms, an energized conductor from a different circuit could fall onto one or more conductors of the deenergized circuit. If the circuit has equipment that is connected to multiple sources of energy, a second or third energy source could be operated to reenergize the deenergized circuit. The equipment could be back-fed through a transformer from a utilization circuit. In some instances, lockout/tagout could be in place and incorrectly implemented. In each instance, a worker performing work on the distribution circuit could be electrocuted due to contact with the unintentionally energized conductor (see color insert, **Figure CP-13**).

Where an exposure of this nature exists, the required electrically safe work condition does not exist. Workers must install safety grounds to control the potential exposure to shock and electrocution. When working on a deenergized conductor, the worker is likely to be a part of the deenergized circuit. To minimize exposure, the worker must install a device that will keep the potential difference between the conductor and surrounding conductive objects to a minimum. Installing an adequately rated conductor from the circuit being worked on and adjacent grounded surfaces will provide low impedance and thereby low potential difference. Installing a grounding cluster ensures that the upstream overcurrent device will see a short circuit and operate in a short time mode.

When current flows through an electrical conductor, a magnetic field is produced in the vicinity of the conductor. The strength of the field depends on the amount of current flow, and the direction of the magnetic field depends on the direction of the current flow in the conductor. The magnetic field associated with each conductor interacts with other magnetic fields and some other metallic components that might be nearby. When the amount of current flow is in the range of the ampacity rating of the conductor, the strength of the associated magnetic field produces little effect. However, when the amount of current approaches the available short-circuit range of a distribution circuit, the magnetic fields can become excessive. When the magnetic fields surrounding each conductor of a three-phase circuit interact with each other, substantial physical forces result. The construction of the ground cluster must be capable of conducting the maximum available fault current in a circuit.

The Right-Hand Rule

The right-hand rule predicts the direction of magnetic lines of force around an electrical wire that is conducting current. The right-hand rule says that if a person's right hand grips the electrical conductor so that the person's thumb is pointing in the direction of the current flow, the fingers on the hand illustrate the direction of the lines of magnetic force (see **Figure 5-8**). As the amount of current increases, the strength of the magnetic lines of force also increases. When current is flowing in multiple conductors that are physically close to each other, the lines of magnetic force interact, resulting in strong physical forces applied to the conductors. If the conductors are a part of a ground cluster, the physical force tends to propel the conductor according to the interaction of the lines of magnetic force.

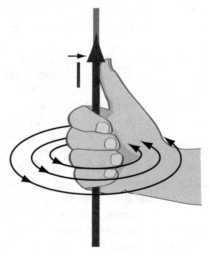

Figure 5-8   The Right-Hand Rule

Performance Requirements

ASTM F855-04, *Standard Specifications for Temporary Protective Grounds to Be Used on Deenergized Electric Power Lines and Equipment*, defines performance requirements for components of these ground clusters and for the complete assembly. Although the possibility exists for employees to assemble adequate protective grounds, the adequacy of the overall construction is not dependable without performing tests defined in the standard. Manufacturers perform tests as a routine part of the manufacturing process and assign fault duty ratings to the completed ground clusters. Only adequately rated ground clusters should be used (see **Figure 5-9**); they are available through electrical distributors.

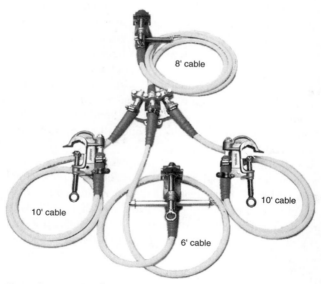

Figure 5-9   Ground Clusters with Four Conductors
Courtesy of Salisbury Electrical Safety LLC

Temporary protective grounds should be installed as close to the work site as possible. A ground cluster should be selected that has a fault duty rating at least as great as the available fault current at the point of the work. Ground clusters must be marked to indicate the rating assigned by the manufacturer.

After determining the necessary fault duty rating and selecting a ground cluster, the conductors, clamps, and connecting points should be visually inspected to ensure that the components of the ground cluster have not been damaged. If any sign of damage to the conductors or clamps is found, another ground cluster should be selected.

Installing temporary ground clusters is one step required to establish an electrically safe work condition (see color insert, **Figure CP-14**). Until the ground cluster has been satisfactorily installed, the worker should consider the conductor to be energized. Wearing appropriate FR protection, the worker should install the ground cluster with a live-line tool or while wearing voltage-rated protective equipment. The first connection should be to an adequately sized grounding conductor. Subsequent connections should be made to each phase conductor. Ground clusters should be removed in reverse order. Remove the connection to the grounded conductor after all connections to a phase conductor are removed.

OSHA requires a zone of equipotential be established for exposed overhead conductors. A zone of equipotential means that all metal components, including conductors, are grounded in such a way that the worker is unlikely to reach outside the zone of protection (see **Figure 5-10**). If the overhead line receives a lightning discharge at another location, the safety grounds would protect the worker, regardless of the location of the strike.

Ground clusters that have been repaired or modified must be tested to ensure that the repaired equipment will

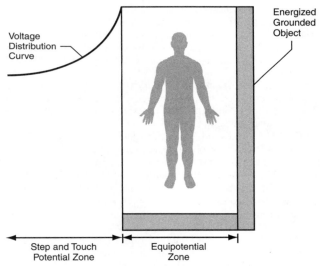

Figure 5-10 Illustration of Zone of Equipotential

pass the standard 30-cycle or 15-cycle voltage-drop values permitted by ASTM F855. Ground clusters should be subjected to the 3-cycle or 15-cycle voltage-drop tests defined in the standard on a regular basis as determined by conditions of use. However, the test interval must not exceed three years.

References

ANSI/ASSE Z87.1, *American National Standard for Occupational and Educational Eye and Face Protection Devices.* New York, NY: American National Standards Institute/American Society of Safety Engineers, 1998.

ANSI/ISA S82.02.01, *Electrical and Electronic Test, Measuring, Controlling, and Related Equipment, General Requirement.* New York, NY: American National Standards Institute/Instrument Society of America, 1999.

ANSI Z89.1, *Requirements for Industrial Head Protection*. New York, NY: American National Standards Institute, 2003.

ANSI Z89.2, *Safety Requirements for Industrial Protective Helmets for Electrical Workers*. (Superseded by ANSI Z89.1 and no longer published, although it remains a reference in 29 CFR 1910.6.)

ASTM F855-04, *Standard Specifications for Temporary Protective Grounds to Be Used on De-energized Electric Power Lines and Equipment*. Conshohocken, PA: American Society of Testing and Materials, 2004.

ASTM F1505, *Standard Specification for Insulated and Insulating Hand Tools*. Conshohocken, PA: American Society of Testing and Materials, 2001.

ASTM F1506, *Standard Performance Specification for Flame-Resistant Textile Materials for Wearing Apparel for Use by Electrical Workers Exposed to Momentary Electric Arc and Related Thermal Hazards*. Conshohocken, PA: American Society of Testing and Materials, 2002.

ASTM F2178-02, *Standard Test Method for Determining the Arc Rating of Face Protective Products*. Conshohocken, PA: American Society of Testing and Materials, 2002.

IEC 61010, *Safety Requirements for Electrical Equipment for Measurement, Control and Laboratory Use*. Geneva, Switzerland: International Electrotechnical Commission, 2002.

NFPA 70E, *Standard for Electrical Safety Requirements for Employee Workplaces*. Quincy, MA: National Fire Protection Association, 2004.

UL 1244, *Standard for Electrical and Electronic Measuring and Testing Equipment. Requirements*. Northbrook, IL: Underwriters Laboratories, 2000.

U.S. Department of Labor. Occupational Safety and Health Administration. OSHA Regulations 29 CFR 1910.132-139, Subpart I, "personal protective equipment." Washington, DC.

U.S. Department of Labor. Occupational Safety and Health Administration. OSHA Regulations 29 CFR 1910.300-399, Subpart S, "electrical." Washington, DC.

Index